Captain James Cook
Seaman and Scientist

Endeavour

Captⁿ. James Cook of the Endeavour.

Captain James Cook
Seaman and Scientist

Bill Finnis

2003/2004 SURREY LIBRARIES	
Askews	
920 COO BIO	

Chaucer Press
London

Published by Chaucer Press
an imprint of the Caxton Publishing Group
20 Bloomsbury Street
London WC1B 3JH

© Bill Finnis (text, black & white sketches, maps and photographs), 2003
© Chaucer Press (this edition and this design), 2003

Bill Finnis has asserted his right to be identified as the author
of this work under the Copyright, Designs and Patents Act 1988.

All rights reserved. No part of this publication may be
reproduced, stored in a retrieval system or transmitted in any
form or by any means, without the permission of
the copyright holders.

A CIP catalogue record for this book is available from the British Library

ISBN 1 904449 14 X

Designed and produced for Chaucer Press
by Savitri Books Ltd

Frontispiece page: *Captain James Cook, painted by William Hodges, the artist who accompanied him on his second voyage to the South Pacific, between 1772 and 1775. In this portrait, Cook wears the undress uniform of a junior captain. This picture is reputed to be Cook's truest likeness.*

Contents

Foreword	12
1. Setting the Scene	13
2. Cook's Early Days	32
3. Preparation for War to the Fall of Quebec	34
4. Surveyor to the King	50
5. Preparations for Cook's First Voyage	56
6. England to Tahiti	63
7. Tahiti to New Zealand	77
8. The Queensland Coast	91
9. Endeavour Strikes a Reef	100
10. Towards Batavia and Home	106
11. Preparations for the Second Voyage	116
12. England to New Zealand	130
13. Pickersgill Harbour to Tahiti	139
14. Tahiti to Queen Charlotte's Sound	145
15. New Zealand to Tahiti	153
16. Tahiti to New Zealand	161
17. New Zealand to England via Cape Horn	175
18. Cook Agrees to Command the Third Voyage	183
19. Preparations for the Third Voyage	193
20. England to Moorea via Cape Town	200
21. Tahiti to Cook's Death in Hawaii	212
22. Epilogue	231
Glossary	237
Bibliography	244
Index	247
Picture Credits	252

Foreword

In an account of this kind, dates present some difficulty. Ashore we are accustomed to the day ending and a new day starting at midnight. At sea the day starts and ends at midday. To that can be added the International Date Line, the line of longitude 180º from Greenwich. When this is crossed on a westerly course, the date is advanced by one day. If the vessel crosses it on an easterly course, the date must be put back one day. There are also the odd occasions when the date in Cook's Journal does not match up with the one mentioned in other accounts by people who were with Cook at the time.

It is easy to see the opportunities for confusion. Whilst I have striven to keep the dates in this biography as calendar dates, I am sure that I must have failed in at least a few instances, and maybe more, for which I apologise. Fortunately, for most of us I suspect, it is not a matter of grave importance.

A few years ago my wife, Betty, and I spent six very enjoyable years, sailing around the world in our thirty-six-foot-long sailing boat, *Didycoy*. The beauty of Australia's Great Barrier Reef, alone, claimed eighteen months, it could so easily have been twice that. Some of the pictures we took during these years have been used in this book, showing that many of the places first seen by James Cook are still as magical today.

In Cooktown – a town of 1700 inhabitants in far north Queensland, founded beside the river where the *Endeavour* found refuge after hitting the reef – we found a small museum, largely devoted to Cook's stay. When the curator heard our Pommie accents he asked us to write a brief account of our voyage so far and to add where we hoped to go on our way home. This account would go into the Museum's log book of visiting yachties. To our delight, the first entry was an account of Cook's visit.

In the rest of Australia, Cooktown has a reputation as an isolated, cyclone-prone, crocodile-infested outpost. To give you some idea of the immensity of the area, the morning we left Cooktown we went ashore to buy some stamps for the letters we wanted to post. Unfortunately the post office was closed. We had to carry those letters some 450 miles to the next post office!

Having sailed into so many of the places Cook visited, it is inevitable that I should have written this account of his life from a seaman's point of view. This means, inevitably, that nautical terms have crept in occasionally. If their meaning is not self-evident, there is a glossary on page 237.

1. Setting the Scene

Much of what follows in this account of Cook's life took place in the Pacific Ocean and this chapter is devoted to setting the scene and taking a brief look at who and what had gone on before him.

I often wonder how the Pacific Ocean got its name. Whilst it is true that for a large part of every year the trade winds hold sway and if you wish to go the way they are blowing, well and good, 'pacific' it is. But the rest of the time some areas of the Pacific can give rise to highly destructive tropical revolving storms, then 'pacific' it certainly is not. Depending where they occur, they have different names: typhoon, hurricane, cyclone and others. But whatever their name, these storms amount to the same thing, ferocious winds that may well exceed 120 knots and the terrifying waves those winds can create in shallow water.

Since winds will be important in this account, it may be helpful to explain that the winds are named for the direction from which they come, not the direction they are blowing towards. Thus an easterly is a west-going wind, a northerly is a south-going wind and so on.

The powerful westerlies in the far south of the ocean do not let up all year and the prudent sailor chooses to run before them, rather than fight them but, even so, the seas they generate in the wrong season, can be highly destructive. Perhaps its very name is an indication of how little was known of this massive ocean when it was named 'Pacific'.

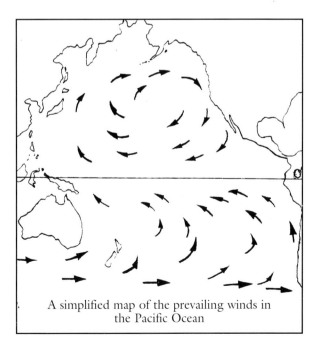

A simplified map of the prevailing winds in the Pacific Ocean

Less dramatic, but equally serious for a sailing vessel, is the wide band of virtually windless sea that spans the equator and is called the Doldrums. It is true that the Doldrums are virtually windless, but it is also an area peppered with the most vicious squalls you could wish to meet in a sailing boat. Many of these squalls grow into tropical revolving storms and move off westwards and away from the equator. It took us forty-two sun-scorched days to cross from Cocos Keeling in the south, to Sri Lanka in the north, where we picked up the wind again, a distance of about 1,500 nautical miles which, in the trade winds, would have taken us about twelve days. At the end of each day, the sky and the sea would become the colour of molten brass or copper. As the sun approached the horizon, it was difficult to believe that the sun was not going to sink into the sea and disappear in a vast cloud of multicoloured steam. Coleridge's *Ancient Mariner* paints the picture to perfection.

Our transit of the Doldrums was made in a well-found yacht with a very healthy crew. Imagine the conditions aboard a much less manageable vessel, that already has a crew depleted by scurvy, and is short of fresh food and water.

Winds in the great oceans run to a well-known circulatory pattern, clockwise in the northern hemisphere and anti-clockwise in the southern hemisphere. These are the trade winds that have been used by sailing vessels ever since their pattern has been understood. Whilst they modify their behaviour from time to time, they have been pretty constant for as long as we have known of their presence. At the time of writing, however, a phenomenon known as *El Niño*, and its sidekick, *La Niña*, are influencing the weather patterns, and not always for the good. At one time their activities were thought to be confined to the Northern Pacific, but they are now known to cause world-wide changes to the oceanic weather patterns and have been doing so at intervals for a very long time.

In these days of electronic navigational instruments that can pinpoint your position with unbelievable accuracy, measure the depth of the water below the vessel, tell you how hard the wind is blowing and in what direction, what course to sail to get to your next point, and so on, you might be forgiven for wondering how seamen of days-gone-by could manage to pilot a vessel in uncharted waters.

I recall sailing with an old Essex yachtsman many years ago, who, when tacking down a river, knew exactly when the depth of water was about to fall, without the aid of an echo sounder. It struck me that this was much better than an echo sounder, as that instrument can only tell you when the depth of water has fallen, *not* when it is about to fall. When I asked him about this talent, he pointed to what he called the 'tide line'. The tide runs at a slower speed in shallow water than it does in deep water and at the junction of the two streams, this difference shows as a small telltale ripple. Water flows faster on the outside of a river bend and this is where you will find the deeper water. On the inside of a bend, the water flows more slowly, so it gives up some of its detritus, making the water on the inside of the bend so much shallower. At the mouth of many rivers a bar builds up, because the river mouth widens, causing the flow of water to slow down and release some of the sediment it has been carrying.

On a sunny day off the Essex coast, the different colours of shoal and deep water can be seen clearly. In the tropics this phenomenon is much more pronounced and a great deal more beautiful. The colours range from the most wonderful cerulean blue you have ever seen, indicating a relatively harmless sandy bottom, to the slate-grey over a reef, that could rip the bottom out of your boat. Deep water, on the other hand, appears as a wonderful rich blue.

White broken water may be caused by an increase in wind speed, or a rock or shoal, just below the surface. The difference is simple to spot because the white water over a shoal remains in the same place.

As for wind direction, you can see it on the surface of the water and, indeed, feel it on your face. You know by the behaviour of the boat if the wind speed has increased or decreased and you can read it from the surface of the water which alters its appearance with every few knots of wind change.

As for the depth of water beneath the boat, a line with a lead weight attached to it and dropped over the side, will tell you that. However, I have to admit that it is an overrated pastime, when you could be sitting in the cockpit with a gin and tonic in your hand, and read the same message on the dial of your echo sounder! But seriously, the seaman of yore was far from helpless and it is a shame that his skills, that put him into such intimate contact with the sea we yachties profess to love, are

Opposite. *An atoll at Cocos Keeling, in the Indian Ocean. This picture was taken from our sailing boat,* Didycoy, *when Betty and I circumnavigated the world a few years ago.*

being supplanted by expensive button-pushing gadgets, that have been known to be fallible or to break down.

In the eighteenth century European governments were becoming conscious of the desirability of finding and claiming new lands. The example of the Spanish and Portuguese looting the gold and silver of the South American Indians and the very profitable trade carried out by the Dutch in the East Indies, encouraged England and France to cast envious eyes on the Pacific.

Some notable voyages had been made, but the Pacific is a vast ocean and they had barely scratched the surface. Until 1790 navigators lacked the ability to fix their position with any reliable degree of precision and so it was almost impossible to return to the same spot of the ocean, except by a mixture of accident and good luck.

In the mid 1700s, there was much talk of a Great Southern Continent, or *Terra Australis Incognita*, an idea whose origins go back a thousand years or more. This was really nothing more than academic theorising and wishful thinking. The argument being that there had to be a landmass in temperate climes to the south of the equator to balance the landmass to the north, and what better place to look for it than the largely unexplored Pacific Ocean. There was not a shred of hard evidence to support this theory, but it did not stop some people from defending the premise quite passionately.

However, for men of power, who wanted more evidence before they became believers, there was always the sneaking feeling that if such a place *did* exist, the possibilities for trade and exploitation would be immense and could not be ignored. The Industrial Revolution was beginning to take shape in England and cheap sources of raw materials, as well as new markets for its products, were glittering prizes indeed. Governments and adventurers alike, looked back on the explorers of old, who had brought back the first tales of discovery from the Pacific regions.

FERDINAND MAGELLAN

Europeans had known very little of the Pacific before Ferdinand Magellan embarked on his three-year voyage in 1519. Magellan, a minor Portuguese nobleman working for Spain, discovered the straits that bear his name and that lead from the Atlantic to the Pacific, between Tierra del Fuego and the southern tip of the South American Continent. He named them the Straits of All Saints but the name was changed to do

Opposite: *The Doldrums – the belt of light winds or calms along the equator*

him the honour he deserves. Magellan believed that the northern coast of Tierra del Fuego was the northern coast of *Terra Australis Incognita*, adding a certain credence to the myth. He failed to investigate this theory, as the purpose of his voyage was to reach the Spice Islands, via the Pacific, rather than crossing the Indian Ocean, which was the normal route. When he emerged from his transit of the straits, he had the great good fortune to pick up the Humbolt Current which took him north, before heading north-west to the Spice Islands and then on to the Philippines, where he was unfortunately killed in a local fracas. Had he had an inkling of the immense distances involved, he may well have had second thoughts about attempting to sail them. As it was, he arrived with more than half his men dead, and the remainder suffering from the effects of starvation and scurvy.

One of the more remarkable things about this voyage is that despite the fact that there are untold thousands of islands and coral reefs scattered around the Pacific, Magellan saw only two minute islets that cannot now be identified. That he didn't pile up on one of the hazards is remarkable; that the two islets cannot be identified is not surprising: it has now been established that, on one occasion at least, his reckoning was over 2,000 miles in error!

It has been suggested that he avoided reefs and islands because he knew they were there. With an error of over 2,000 miles in his reckoning, it seems highly unlikely that he could have avoided an unseen obstruction in the sea, by navigating around it!

I suppose, people like Magellan had some idea of what they were sailing into, but much of that 'knowledge' would have belonged to the realms of myth, speculation and rumour.

The courage required of someone with a ship that would be considered a death trap today, setting out on a voyage of that kind, with so little knowledge of what lay ahead, must have been immense. To compare it with space travel today is to do it a total injustice. Astronauts have all the information they need and a vast team of highly trained specialists to come to their aid, should they run into difficulty, not to

Opposite: *This detail of an early 'World Chart', drawn by Pierre Descellier in 1550, shows the kind of scenes and animals expected to be found in the Great Southern Continent. Since antiquity, it had been thought that such landmass was necessary to counterbalance the weight of the land in the northern Hemisphere. Many still believed in the existence of this mythical landmass when Cook left for his first voyage, aboard* Endeavour.

mention a rescue craft on standby, in case the worst should happen. Of course, and in spite of all these safety nets, things did go horribly wrong on two space voyages, and this perhaps says something about the nature of adventure and the discovery of new frontiers. However much man tries to control his environment, terrible accidents can and do happen. But in the days of sail, the chances of a seaman making it back alive, were hardly fifty percent.

As early as 1530 French cartographers had produced charts that depicted Australia with a shape that was recognisably that of Australia. The east and southeast coasts were shown as a rather sketchy outline, but the rest of the coastline, errors and all, was undoubtedly Australia.

It is believed that the material used to produce these charts was stolen from the Portuguese, suggesting that they must have gone there, well before 1530. Did Magellan's expedition bring enough information back to make these charts possible, or were there others of whom we have no knowledge? It is not surprising, that what is now the Queensland coast, was not shown in the same detail as other areas of the coastline. The Great Barrier Reef extends from a little north of what is now Brisbane, for about 1,000 miles to Cape York, and on beyond to New Guinea, stretching out into the Coral Sea to distances of forty to 150 miles. In places, the reef comes to within a mile or less of the Queensland coast and is an area that would have proved deadly for the unhandy vessels of the day, even if they had possessed the detailed charts we have today.

ANDRÉS DE URDANETA

Following Magellan's discovery of the Philippines, a measure of trade developed between those islands and the Spanish-held ports on the West Coast of Mexico. The outward-bound passage would follow the west-going trade winds, just north of the equator. The return passage was rather more difficult and this limited the volume of trade until 1565, when Andrés de Urdaneta found the east-going winds in the northern part of the North Pacific, which would take him back to the west coast of Mexico.

ÁLVARO DE MENDAÑA and SARMIENTO DE GAMBOA

Not long after Urdaneta's voyage Álvaro de Mendaña and Sarmiento de Gamboa set out from Peru in search of the 'Islands of Gold', which, according to Inca legends, were to be found some 600 leagues (probably 2,400 nautical miles) to the west. They failed to find them, even though they sailed three times that distance. They did, however, find what are now known as the Solomon Islands.

Thirty years on, Mendaña attempted to return with the Portuguese, Pedro Fernández de Quirós, as his pilot and a ship-load of would-be colonists. They found the Marquesas but carried on in search of the Solomons. Mendaña died and the surviving colonists showed a decided lack of interest in the project, forcing de Quirós to alter course to pass to the north of New Guinea, and thence to Manila in the Philippines. From there, they fell in with the east-going winds that took them back to Mexico and then it was a matter of coasting down to their starting point in Peru.

SIR FRANCIS DRAKE

For much of the sixteenth century relations between England and Spain were uneasy, at times teetering on the brink of war. Spain was well established in South America and made sure that English merchants were excluded from trading with what is now Latin America. At the same time the Pacific was also very much a Spanish domain. The Spaniards had established a number of settlements along the Pacific coasts of South and Central America from which they carried on a regular trade with the Philippines and the Mariana Islands, exchanging looted gold from Peru and Mexico for the silks, spices and other valuables from the East Indies. Their trading vessels would then return to the west coast of America and the goods were carried overland to the Caribbean coast where, along with large quantities of gold and silver, they were loaded aboard Spanish galleons for transit to Spain.

English investors and merchants became increasingly frustrated and resentful of the Spanish monopoly in these areas and attempted to get Queen Elizabeth I to adopt a more belligerent attitude towards the Spaniards. For a long time the Queen was reluctant to take steps that could lead to war, but covertly encouraged those buccaneers who were willing to attack the Spanish treasure ships and relieve them of their precious cargoes. For a number of years, Francis Drake had been a highly successful and prominent member of this band of buccaneers.

It was not until 1577, when relations between Spain and Britain were worse than ever, that Queen Elizabeth gave her approval for a voyage through the Straits of Magellan and on into the Pacific. Her grip on the purse strings was as firm as ever, and little or no State money was forthcoming to help the venture.

Francis Drake was appointed to command the expedition. His instructions required him to explore the Atlantic and Pacific shores of South America, with a view to finding areas not yet claimed and colonised by the Spanish or the Portuguese. Beyond that, Drake's instructions were ambivalent and open to a variety of interpretations, which undoubtedly suited him. The gentlemen who sailed with him, were under the impression that they had agreed to a voyage of exploration, to search for *Terra Australis* and the Northwest Passage, all with a view to trading in the Pacific. Drake was more interested in finding and plundering Spanish shipping and, since Drake was the commander of the expedition and, I suspect, a far more forceful character than 'the gentlemen', that is exactly what he did.

By November 1577 Drake aboard the *Pelican*, (it was renamed the *Golden Hind* later in the voyage), was on his way south with four other vessels. They entered the Pacific via the Straits of Magellan and negotiated them with none of the usual difficulties. The expedition paid for this easy passage by running into a severe storm, almost as soon as they emerged into the Pacific. The storm drove them south and east of Tierra del Fuego, until they were out of sight of land, thus proving that, at the very least, it was not *Terra Australis*. Indeed, because of the nature of the winds and seas, Drake felt confident that there was a route from one ocean to the other to the south of Tierra del Fuego. His instincts were correct and today the area is called Drake's Passage.

The *Golden Hind* was now the only surviving ship. Earlier on, Drake had abandoned one vessel as unseaworthy, another had sank with all hands in the middle of a storm and within sight of Drake's barque. Contact was lost with the remaining two vessels.

Once the weather allowed, Drake sailed northwards, raiding and stealing as he went, eventually capturing the *Cacafuego*, a treasure ship. Her cargo of gold, silver and precious stones was transferred to the *Golden Hind*. It was time to go home. To the south, the Spaniards were alerted to the presence of Drake's ship and were almost certainly waiting to attack him and reclaim the treasure he had stolen from them. Drake continued as far north as the present-day city of San Francisco, then turned to the west and crossed the north Pacific.

He called in at some islands, believed to be the Carolines, where he picked up some fresh food and moved on fairly quickly, as the locals were not particularly friendly. From these islands *Golden Hind* continued westwards to the Spice Islands and took on as much spice of one kind and another as could be loaded on board. Shortly after leaving, *Golden Hind* struck a reef, just off the coast of the Celebes, and was pinned there for some days until the wind shifted and they were able to get her off, but not before jettisoning much of the spice bought in the Celebes. From there, it was Cape Town and home again to England. Much as one can admire Drake's tenacity in completing the voyage, he contributed little to the knowledge of the world, except for his comments on the area around Cape Horn.

PEDRO FERNÁNDEZ DE QUIRÓS and LUIS DE TORRES

1605 saw Quirós sailing westwards again, with a few ships and in company with Luis de Torres. Quirós believed that the Solomon Islands were part of the Great Southern Continent and his mind was set on finding it. This time Quirós's course took him a little further to the south and, instead of reaching the Solomons, he found what is now known as the New Hebrides, or Vanuatu. He was convinced that he had found the fabled *Terra Australis* and named the islands *Australia del Espiritu Santo*, claiming them for the glory of Spain and that of the Holly Ghost. Quirós then, sailed back to America, picking up the winds Urdaneta had found.

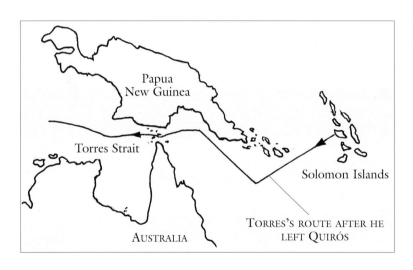

When Quirós headed off east, Torres went his own way. He was far from convinced that they had found the Great Southern Continent and he sailed a further 400 miles southwest. Had he continued for a further 1,000 miles or so, he might well have stumbled upon Australia, but chances are that he would have come to grief on the Great Barrier Reef in the Coral Sea, or did he perhaps know what he was doing? When he had completed his 400-mile stretch on a southwesterly course, he turned to the northwest and, a little later, to west, northwest – courses which took him safely clear off the south-eastern tip of New Guinea and parallel with its southern shore. Such a succession of courses seemed to have been taken from choice rather than whim. Torres maintained his course, passing between Cape York, at the north-east corner of Australia, and the southern coast of New Guinea – the passage that is now known as the Torres Strait. Somehow, one can't help thinking that he knew what to expect. Here he is, with a vessel that would have had great difficulty in beating back against the wind and which lacked the ability to manoeuvre easily in tight situations. He sailed before the wind into an ever-narrowing channel, beset by coral reefs, breaking through into the Arafura Sea and then the Timor Sea, as if he knew where he was going, but this theory will have to remain speculative.

WILHEM CORNELISZEN SCHOUTEN and ISAAC LE MAIRE

Two Dutchmen, Wilhem Corneliszen Schouten and Isaac le Maire, are credited with pioneering the route around Cape Horn in 1615. They were not enamoured of the Straits of Magellan's route through to the Pacific and had probably heard of Drake's theory on the subject of an easier route to the south of Tierra del Fuego. Whether they were influenced by Drake or not, they sought an alternative passage into the Pacific and found it by going south of the southernmost point of the South American Continent. They named it Cape Hoorn, after the town of that name in Holland. It has since been Anglicised to 'Cape Horn'.

Writers and film makers have given the Horn an evil reputation and, in the wrong season, it deserves it. In the right season it can be relatively easy, indeed the last yachtsman I spoke to on the subject, said that he'd had to motor all the way round for lack of wind! We also know that Captain Cook was becalmed for a few days on his first rounding of the Horn.

Schouten and Le Maire lost one of their ships, but made their way in the remaining vessel to the north of the Solomons, Australia and the New Hebrides, before reaching Java. Their fellow countrymen refused to believe them when they claimed to have found a new route into the Pacific and jailed them for making fraudulent claims. I suspect they probably coined the phrase: 'I don't know why we bothered!'

A little later, another enterprising Dutchman reasoned that if there was a strong west wind to the south of the Cape of Good Hope, perhaps it would persist long enough to carry him across the South Indian Ocean to the longitude of the Spice Islands. Once there, he could turn north, with a reasonable chance of meeting up with the islands he sought. He was successful and the route he pioneered was used increasingly, as it presented a number of advantages over the slower route, further north in the Indian Ocean, but it did require strongly built ships and good seamen. The Dutch had both.

The new route required the navigator to know what longitude he had reached and the problem of determining a ship's longitude had yet to be solved. As a result, many Dutch ships bound for the islands to the north, must have overshot their longitude and foundered on the west coast of Australia. Certainly the Dutch of those days called Australia 'New Holland'. It is highly likely that many Dutch navigators would have placed themselves on a latitude that would take them to the west coast of Australia and then turn north to their destination. It was possible to establish a ship's latitude at this time, but it was to be many more years before the problem of establishing longitude from a ship's deck was solved. Fortunately for British interests, the Dutch felt that New Holland held no attractions for them, perhaps they were doing too well trading with places like Java, Timor, Bali, Indonesia, and many others, to be bothered to explore what they saw as a barren and inhospitable land.

CAPTAIN GEORGE ANSON

George Anson left England for the Pacific in 1740. England was at war with Spain and the aim of the voyage was to harass Spanish commercial shipping and parts of Spain's Empire in the Pacific. Anson had four ships of war and a number of store ships, which sounds like a strong force. In fact, they had been so badly provisioned and inadequately supplied with gear by the incompetent and corrupt dockyards back in England, that it was a disaster waiting to happen.

Anson attempted to take his squadron round the Horn, in the face of vicious storms. Two vessels failed to weather the Cape and a third was wrecked on a lee shore, on the southern Chilean coast. The crew of *Centurion* were going down so fast with scurvy and the illnesses they had brought aboard with them, that Anson soon had barely enough men to run his ship.

Captain Anson's ship was the only one left and he carried on regardless into the Pacific. In 1743 he had the good fortune to capture a large Spanish galleon, the *Covadonga*, which was filled with gold and other treasure. This catch enriched the coffers of England and made Anson a wealthy man. The crew fared less well: of the 2,000 men who had sailed out with Anson, when he started the voyage, 1,800 died, many from scurvy.

The real value of the expedition came from Anson's experience of what must have been a nightmare voyage. He was raised to the peerage and, in 1745, was appointed to the Admiralty, where he strove to improve matters until his death in 1762. For much of this time, he served as First Lord of the Admiralty. Anson got rid of the dead wood and corruption in the administration and tried to improve the dockyards. It was made clear to incompetent officers that their ways would no longer be tolerated. Many were not given another ship and kept on half pay, while others were replaced by younger men. Discipline on the lower deck was also tightened. Anson had the ability to choose men of valour and to put officers in positions where they could exert their influence to the benefit of the service. His work restored the Royal Navy's efficiency and its morale. It was James Cook's good fortune to serve in the Royal Navy at a time when the results of Anson's labours were coming to fruition.

LOUIS ANTOINE DE BOUGAINVILLE

Bougainville, a Frenchman, was commissioned by his government to sail around the world. He entered the Pacific, shortly before Cook. Bougainville spent some time in Tahiti before Cook's arrival and then moved on, stopping briefly at Samoa, passing close to the New Hebrides and the eastern fringe of the Great Barrier Reef, without sighting Australia. By this time, his crew were suffering from scurvy and his ship badly needed a refit. He put into Buru in the Moluccas, to carry out the work necessary to put his vessel in order and to refresh his crew before sailing for home.

Whilst he roamed far and wide, he left few charts, his position fixing was rather sketchy and lacking in precision. An island to the northern end of the Solomons bears his name, as does the strait between it and the Island of Choiseul, immediately to the south-west of Bougainville. He is also remembered as the namesake of a rather pretty tropical flowering shrub: bougainvillea. Cook was very critical of his position fixing and claims that Bougainville's crew were responsible for introducing venereal disease to some of the Pacific Islands. Whilst it is not the sort of thing that can be pinned down with certainty, the evidence suggests that Cook may well have been correct.

WILLIAM DAMPIER

Dampier was born in 1652. His father farmed in Somerset and, like Cook, he was away to sea as soon as he was old enough, but that is where the similarity ends. Unlike Cook, he tried his hand at a succession of seafaring occupations. His first experience was in the Newfoundland trade, taking salt out and bringing cargoes of dry cod back to England. The work and the weather proved to be too tough and cold for him and he tried a voyage to the East Indies. He quickly realised that if he didn't die of scurvy en route, he might well do so of some dreadful fever after arriving in the East Indies. At the age of twenty, he entered the Royal Navy as a seaman. After his time in the Navy, he tried a spell of buccaneering in the West Indies and then in the Pacific, where he travelled extensively, until his crew, fed up with all this seagoing with no profit in sight, put him ashore on the Nicobar Islands. After a further succession of jobs, he managed to get a passage back to England in 1691 and he turned his hand to writing. Dampier was intelligent and observant and had kept a meticulous log throughout his seafaring ventures. His first book, *A New Voyage Around The World*, sold four editions in two years and was responsible for awakening interest in the Pacific. Two more books followed, one of which was entitled: *A Discourse on Winds*, a truly invaluable treatise on the wind patterns he had observed around the world.

The Admiralty had decided to organise a single ship expedition to explore the Pacific, especially to find and claim *Terra Australis Incognita*, should it exist. At this time the Admiralty were prepared to allow outsiders to command their ships and Dampier was offered the post. On the face of it, he seemed a sensible choice but, in spite of his many nautical skills, he totally lacked the two essential qualities: leadership and the power of command. The ship he was given was in a dreadful condition, the planks were rotten and the rigging no better. His acceptance of a vessel in such a ripe condition must say something about the man. From the start, the crew were ready to mutiny and his officers resented having to serve under a captain who had been a buccaneer and whose only prior service in the Royal Navy had been as an ordinary seaman.

How Dampier managed to sail that rotten hulk to New Guinea, then turn for home and get as far as Ascension Island in the South Atlantic, before the ship sank beneath him, is a mystery. She sank when the shipwright attempted to repair a leak and the timber crumbled under his hands. The ship's company was able to row to Ascension, from where they were picked up a few weeks later and eventually taken home by a squadron of English ships.

Dampier was court-martialled by a totally unsympathetic, some might say an utterly biased group of Royal Naval officers. He was sentenced to forfeit the pay he had earned and was declared a person unfit to command a naval vessel ever again. This surely must be rated as a harsh sentence, passed by representatives of a Service that was responsible for sending such an ill-equipped vessel to sea. For Dampier, it was back to buccaneering and writing, until his death in 1714.

COMMODORE, THE HON JOHN BYRON

In June 1764 Commodore Byron was despatched to find lands that were believed to exist in the South Atlantic. He was instructed to visit the Cape of Good Hope and then continue south to latitude 53º, where he should turn to the west, seeking the Virgin Islands, that had been claimed by Hawkins many years before. When this was completed, he was to find Pepy's Island and after that, move on to the islands that are now known as the Falklands. He was to take formal possession of these islands for Britain and to carry out surveys. When he had completed those tasks he was to pass into the Pacific, heading northwards, until he found New Albion, which had been reported by Drake to lie at 38º north. He was to reassert British sovereignty over New Albion and then go further north in an attempt to find the Northwest Passage.

The existence of a Northwest Passage, across the top of North America, that would allow ships to sail from the Atlantic to the Pacific, had been a long-held belief. Like the *Terra Australis Incognita* theory, it was not supported by fact. In 1516 William Baffin had explored Hudson Bay in search of a possible Northwest

Passage and had reported that its existence was unlikely. In spite of this, the pressure was on to find it, this time from the Pacific side. This was to be a massive enterprise and Byron was not the man for the job.

Disregarding the instruction to go to the Cape of Good Hope, Byron sailed straight to the Falklands and spent some time searching for Pepy's Island, finally deciding that the Falkland Islands and Pepy's Island were one and the same.

On 11th February 1765 he started the transit of the Magellan Straits which took the two ships, *Dolphin* and *Tamar*, three months to complete, by which time Byron judged his ships unfit to continue with the tasks he had been set. He laid a course for the East Indies, where he expected to be able to refit his vessels, before tackling the passage home. If it seems strange that Byron was prepared to sail so far, before turning for home, it must be understood that his ships, in common with most vessels of his day, needed the wind to be somewhere behind them to be able to sail at all. Even the skipper of a modern yacht, designed to sail within some 45° from the wind, would normally prefer to sail *away* from the wind, rather than sail into it. Progress would be faster, and certainly more comfortable, which also means less strain on the ship's gear. Byron got back to the Thames in May 1766, having achieved very little of value to future seamen.

CAPTAIN SAMUEL WALLIS and CAPTAIN PHILIP CARTERET

As soon as Byron had relinquished command of the *Dolphin*, a refit was started to make her ready for another voyage into the Pacific Ocean, this time under the command of Captain Samuel Wallis, and in the company of a sloop, named *Swallow*, commanded by Captain Philip Carteret. (At this time, in the Royal Navy, a sloop was a small vessel, regardless of its rig.) Their instructions were to enter the Pacific, via Cape Horn or the Straits of Magellan, and then to proceed westwards, in the most southerly latitudes they could maintain, in search of the Southern Continent. They were to go as far as New Zealand and, if they failed to find the Southern Continent, were to explore further north, before returning to England.

Following Byron's advice, the two ships entered the Straits of Magellan on 17th December 1766, mid-summer in the Southern Hemisphere. On 11th April and 350 miles later, Wallis finally emerged into the Pacific. He was not safe until he was far enough from the rock-bound coast to be able to work his way offshore, should the weather turn bad. Sea room, distance from a coast line, is invaluable in heavy weather. Today's long-distance yachtsmen have coined the phrase: 'If I can see it, I am too close' and it sums up the situation neatly.

Wallis had suffered considerable delay through the straits, having to wait for *Swallow* which sailed very badly. The Square-rigged ships of the day could sail no closer to the wind than about 70º and to advance into the wind it would be necessary to tack back and fore, across the wind – sailing six miles to gain two into the wind. In heavy weather, ships of this kind, would be driven sideways away from the wind so that the 70º could well become 90º or more. Even so, Wallis was considering the possibility of going back into the straits to do what he could for her, when the wind shifted and fog would have made it impossible to beat out of the straits, should he have sailed in again. Prudent seamanship dictated that he should move offshore.

As for *Swallow*, it was to be another week before she cleared the Straits of Magellan. Now that they were separated, both captains attempted to carry out their instructions. Carteret carried on, in his small vessel that had been badly battered and was ill-equipped for ocean exploration. He completed the circumnavigation and carried out a creditable amount of exploration, but as he did not reach England until twelve months after Cook had left on his first voyage, the latter could not benefit from the knowledge gained by Carteret.

Wallis continued westwards in the higher latitudes of the southern ocean and had a very hard time. The delay the two ships had suffered had taken them into the southern winter. By staying so far south, not only were the conditions bad in every respect, but Wallis was attempting to sail into the teeth of very strong winds, in a vessel not designed to do that. He fought his way westwards for nineteen days, against seas that threatened to shake the masts out of *Dolphin*. There was not a dry place in the ship and the surgeon warned Wallis that the crew were going down with fevers and scurvy at a rate that would soon leave him with too few fit men to run the ship. Wallis ordered the course to be altered to the north-west to seek places where his crew could recover. He found a number of islands that had not been sighted before, notably the Tuamotus, a large area of beautiful atolls, and Tahiti, which he named King George the Third Island. Before leaving the area, he visited other islands of the Society Group and then turned for home via the Ladrones (now the Marianas), Batavia (now Jakarta) and the Cape of Good Hope.

Wallis arrived back in England one week before Cook received his commission as a Lieutenant on 26th May 1768. The new knowledge Wallis brought back from the Pacific strongly influenced the instructions that were to be given to Cook.

ALEXANDER DALRYMPLE
So much for the men who had been there. Now we must look at the man, who

perhaps best represents those who believed passionately in the existence of the Great Southern Continent, but had never been there to look for it. He was one of that breed of scholars who gathers all the information that supports their thesis, and disregards that which does not. Alexander Dalrymple had some limited sea experience, but was a member of the Royal Society, an astronomer, a cartographer and a surveyor of some note.

The planet Venus was due to cross the face of the sun in 1769. Edmund Halley, the great Astronomer Royal of comet fame, had read a paper before the Royal Society, in which he advanced the theory that if a number of observations of this event were to be taken at widely scattered points around the earth, it would make it possible to calculate the distance to the sun with some accuracy. The ability to make this measurement had so far eluded mathematicians and astronomers and it was of considerable importance to them.

The Royal Society was keen to organise the observation of the Transit of Venus from Tahiti and petitioned the King for financial support. George III told the Admiralty that the Royal Navy should supply a suitable ship for this undertaking. It seems that this was the first the Navy had heard of the project. Certainly Cook knew nothing of it, he had been on leave until 4th March 1768, when he reported to the Admiralty and then set about preparing the *Grenville* for the coming summer season's survey work in North American waters.

The Admiralty still hankered after clearing up the question of the Southern Continent and was looking for a ship to undertake that task, when the King's request on behalf of the Royal Society, arrived. The Navy realised they could combine the two expeditions and all this led to Cook's first Pacific voyage.

Barquantine

Overleaf: *Heading toward home.*

2. Cook's Early Days

October 1728 to Early 1755

James Cook was born on 27th October 1728 at Marton, a small village in Cleveland, Yorkshire. He was the second child of a large family, few of whom survived. Living conditions and the limited state of medical knowledge made for large families and few survivors at all levels of society. His father, also named James, was from Scotland and worked as a farm labourer. His mother, Grace, was a Yorkshire lass. Cook's father was a good, intelligent worker and his employer, Mr Thomas Skottowe, promoted him to manage Aireyholme Farm at Great Ayton.

James did odd jobs for a Miss Mary Walker, an old lady of the village. His intelligence obviously shone through because she offered to teach him to read. At the same time, Mr Skottowe paid for him to attend the village school, where he kept very much to himself. Schooling ended at an early age for the children of the working classes and, aged twelve, James joined his father at work on Aireyholme Farm, where he stayed until he was seventeen. He was then apprenticed to a Mr William Sanderson, a haberdasher and grocer, in the small fishing port of Staithes, about ten miles north of Whitby, which was also notorious as a smuggling port. This is probably when the young man first felt attracted to the sea.

Shopkeeping quickly became distasteful to Cook and the sea more and more attractive. Eventually, he prevailed upon his employer to release him from his apprenticeship. With his parents' consent, Mr Sanderson took him the ten miles to Whitby, where he was apprenticed to the Walker brothers – Quaker shipowners of Grape Lane, Whitby – in the July of 1746. James Cook was going on for eighteen and it was unusual for apprentices to be taken on at that age. Normally, apprentices started three or four years earlier and it undoubtedly took a deal of special pleading on the part of his sponsors to get Cook apprenticed.

Looking at Cook's early days, it is obvious that he was blessed with the support of a succession of kindly, understanding and discerning people who helped him on his way. Having said that, he must have shown considerable, if not exceptional, promise both in intelligence and character, for them to have made the necessary efforts on his behalf.

Right. The typical squat shape of a Whitby collier. Endeavour, *which started life as the* Earl of Pembroke, *was one of these serviceable Whitby 'cats', plying the waters of the North Sea.*

Cook joined a number of other apprentices aboard the collier *Freelove*, a ship of 450 tons engaged in the coal trade between Newcastle and London. He sailed in her for two seasons, spending the winter months in the home of his master, Mr Walker, as was the custom for apprentices. These winter months, whilst the ship was laid up, were not wasted. He spent them in study, helped and encouraged by Mr Walker, with whom he developed a lifelong friendship.

Early in 1748 the Walkers purchased a new ship of 600 tons, called the *Three Brothers* and Cook was put to work on her rigging. When the *Three Brothers* was ready for sea, he made two voyages in her on the Newcastle-to-London coal run and one to Norway, then the Government chartered her to convey troops from Liverpool to Dublin.

In the summer of that year, his apprenticeship completed, he transferred to the *Mary* which was skippered by a Mr Gaskin, a relative of the Walkers, and served some time with him. For a while, Cook shifted his berth to a ship from Stockton but he returned to Whitby in 1752. The Walkers offered him a post as Mate aboard the *Friendship*, he accepted and remained with her until the spring of 1755. Most of the work was in the familiar coal trade but, from time to time, there would be a trip to the Baltic, which enlarged Cook's experience.

Early in 1755 the command of the *Friendship* fell vacant and it was offered to Cook. To everyone's surprise, he turned it down. Command of a trading vessel offered opportunities for private ventures and personal enrichment that few would refuse. To suggest that Cook was not interested in money would be wrong, but it was far from being a prime consideration.

His was a mind that soaked up new ideas and learning, but he then needed the opportunities to put that learning into practice. Cook could see that the Royal Navy would offer him a far more varied experience than commerce ever could. The climate in the country was ripe for enlarged horizons and greater aspirations. Commodore Anson had just returned from his successful brushes with Spanish ships and was now a wealthy man, who had come back to a hero's welcome.

James Cook saw the Navy as his chance to escape the monotony of a life devoted to the transport of coal, with an occasional foray into the Baltic, and exchange it for a life with much wider horizons. This said, at the time, he probably had little idea how far this decision would take him.

3. Preparation for War to the Fall of Quebec

Summer 1755 – October 1759

In the summer of 1755 Britain was arming herself for yet another war with the old enemy, France. The French were setting up colonies in North America and this was not to be tolerated. They were also encroaching on British interests in India, and that too had to be dealt with. With both these interests so far away, it was necessary to transport men and materiel safely across the sea and the Navy was building ships of war at a fast pace to do just that. Cook had joined at a good time from his point of view: the Royal Navy had three important targets in sight: to dominate the Indian Ocean, to blockade France and to attack the French in North America.

The first of the English settlers had landed on the eastern shores of America and, in the early days, the Indians, the native population, had generally been helpful to them, indeed it is unlikely that the settlers would have survived the first few winters without their help. Sadly these good relations did not last. As the settlers attempted to enlarge their areas of settlement, goodwill evaporated, as either side defended itself against the other, until mistrust and hatred became the order of the day.

Meanwhile, the French, who had landed on the eastern shores of what is now Canada, had sailed across the Great Lakes and made their way south, to the west of the English colonists. In the course of their southward march, they had reached the Mississippi and the immense fertile plains through which it flowed. At the same time, the English were penetrating the territory to the north, through what is now western Pennsylvania and New England.

This northern expansion was being attacked by Indians on a regular basis, whilst the French seemed able to get on better with the local populations. Not unnaturally, the English believed, rightly or wrongly, that the French were encouraging the raids on English settlers. This, allied to the French penetrating and settling in areas that the English would rather have had for themselves, was enough to provoke armed conflict between them.

In 1755, aged twenty-seven, James Cook joined the Royal Navy as an able seaman. This was quite a decision for a man, who had recently been offered and turned down, his first command. We'll never know exactly what prompted him. His only reported comment on the subject was that: 'he had a mind to try his fortune that way.' Be it as it may, the Royal Navy lost no time in bundling James Cook, their newest recruit, into a Portsmouth-bound coach His fellow travellers were business men who spent much of the time discussing ways and means of making a quick penny out of the forthcoming war. Cook sat and doubtless pondered on the whys and wherefores of what was in store for him.

He had been drafted to the *Eagle*, a sixty-gun ship that was being fitted out at Portsmouth. He found her to be bigger than anything he had worked on before – four times the size of *Friendship*. Everything about her was enormous by comparison with the vessels he had worked on before, but despite the difference in size, the rigging and spars were really only *Friendship* writ large.

When Cook joined her, the disorganisation aboard *Eagle* was total. Wherever he looked a state of chaos reigned with little of value being achieved. Cook reported to the elderly Master and was set to work on the rigging. There was a small group of seamen whose skills were more than adequate for the task in hand, but they lacked someone to co-ordinate their efforts. Quite soon, they were accepting Cook as their natural leader, and the work began to progress as it should.

Not only did other seamen appreciate James Cook's skills and leadership, but so too did the Master and the officers in charge, so much so that Cook was quickly promoted to Master's Mate. In his day, a Master's Mate was the equivalent of a present day Petty Officer. A big ship would have several Master's Mates to help the Master run the ship. Between them, they were responsible for about everything that happened, short of the tactical and strategic use of the vessel.

Royal Navy officers came up largely through a hierarchical system, but, on the other hand, high birth was usually enough to secure a commission in the Service. Merit alone was not always enough, but if allied to a measure of patronage, then advancement through the commissioned ranks was highly likely. To break through the barrier from the lower deck, as Cook was to do later in his career, although possible, was a highly unusual achievement. The best most lower-deck men could aspire to was to become a Master, which was an honourable career in itself, but one with a decided ceiling to it. In the Royal Navy of Cook's day, a Master was a warrant officer, somewhere between a petty officer and a commissioned officer. The rank was a hangover from earlier days, when a Master was responsible for the maintenance and running of a vessel, but did so at the behest of the senior military man in charge of the soldiers on board. Those who led the fighting men came from the higher reaches of society and the gulf between them and a ship's Master, was both deep and wide.

By the time Cook joined the Navy, the Captain of a vessel undertook the navigation and pilotage, with the Master playing a supporting role. The Master was still responsible for the maintenance and the working of the maze of ropes that supported the masts and controlled the sails, and a great deal more besides. The rank of Master was an important position in one of His Majesty's Ships and, ceiling or no ceiling, for a man of Cook's calibre, it was a great step forward from being a Captain engaged in the coal trade for the rest of his life.

As soon as *Eagle* was ready for service, Captain Hamer came aboard and took up station to blockade the stretch of water between Land's End and Cape Clear, the southernmost point of Ireland. As *Eagle* made her first approach to the Irish coast, she encountered a severe storm off the Old Head of Kinsale, and suffered so much damage, that it was decided to put into Plymouth for repairs. Captain Hamer lingered in Plymouth for so long that the Admiralty showed its displeasure by replacing him with Captain Hugh Palliser.

Palliser proved to be a very different breed of commanding officer and this was to be greatly to Cook's advantage. Captain Palliser was to become an Admiral, a Governor of Newfoundland, a Lord of the Admiralty and an influential patron and friend of James Cook.

Within weeks of taking up his command, Captain Palliser realised that Cook was a man of above average intelligence and competence and had him promoted to Bosun. Whilst a Bosun was still a petty officer, the rank made him an important and senior petty officer, a definite step towards the position of Master.

Under Palliser's command, *Eagle* became a different ship. There was a sense of urgency to prepare it for the likelihood of action and plenty of hard patrolling, summer and winter, in the Western Approaches.

James Cook quickly found that blockade work was very much more onerous than simply shifting cargoes of coal around the coast. No matter what the weather, the blockade had to be maintained, and maintained it was for months on end. Although for much of the time they were in sight of the shore, this did not mean that they escaped the scourge of scurvy. After only a few months on blockade duties, *Eagle* landed 130 men at Plymouth, all seriously ill with scurvy, having earlier on buried twenty-two men at sea. This was a new experience for Cook. Prior to this, his passages had been counted in days, or a week or two at most, and scurvy never had time to take hold. Now his ship was expected to keep to the sea for weeks and even months on end and the consequences for the health of the crew could be dire.

Whole ship's companies would fall foul to this malady. Yet even losses on this scale were accepted as an inevitable consequence of lengthy ocean voyages. However, there were a few, who had different ideas. Fifteen years before cook was to leave England on his first voyage to the Pacific, a Doctor Lind, who had served in the Royal Navy, where he had seen the results of scurvy at first hand, published a paper in which he stated that scurvy was caused by a lack of fresh fruit, vegetables and greens. His findings were the same as those of Johann Friedrich Bachstrom, a Polish doctor. Dr Bachstrom's treatise was written in Latin and it was not until Lind translated it that his findings became more widely known. Dr Bachstrom had noted the symptoms of scurvy among the inhabitants of towns that had undergone a siege. He too had concluded that the disease was caused by their diet. This observation disposed of the long-held belief that 'emanations arising from the putrefaction and corruption of the sea' were to blame for scurvy.

Lind had also noticed that, although the people of Newfoundland had no access to fresh fruit and vegetables during the harsh winter months, they stayed free of scurvy. He attributed this correctly to a beverage made from the needles of the spruce tree which these people prepared and drank during winter.

Scurvy was a most distressing condition. The symptoms were easy enough to detect and followed a well-known pattern. After eight or nine weeks of subsisting on dry rations, the men first acquired swollen and bleeding gums, the teeth eventually falling out. The connective tissue in the victims' bodies deteriorated, causing bruises to appear for no apparent reason, the limbs swelled and wounds failed to heal. Blood would seep into the joints and under the skin, causing agonising pain and stiffness. Creeping anaemia would totally sap the strength of the sufferers, who often died bloated and blackened after several weeks of intense pain. We now know that scurvy is due to a lack of vitamin C which is found in fresh fruit and vegetables and these were almost totally lacking in the diet of ocean-going seamen.

Cook was horrified by the human wastage and studied all the available information on the subject. Later on, when he had his own command, he was able to put his remedial measures into practice. He would ensure that a constant supply of fresh food, and anti-scorbutic agents, were kept on board, achieving a remarkable degree of success. Of course, in pre-refrigeration days these apparently simple measures were anything but simple. The situation was made worse by the latent corruption operating in the dockyards back in England, and among those whose duty was to supply food to the King's ships, to the detriment of the seamen. They supplied inferior quality meat, which was stowed in barrels and salted; the cheapest cheeses available, often infested with worms; ship's biscuits, usually riddled with weevils. In addition, the men were issued with a ration of dried peas and oatmeal. The peas were probably the only item of food that had the least hope of keeping scurvy at bay, but obviously it was not consumed in sufficient quantities to be really effective.

A very poor and monotonous diet was not the only cross the seamen had to bear: the living conditions for the crew aboard most of the King's ships were quite dreadful. The number of men needed to man a warship was such that the accommodation was overcrowded beyond belief. As well as the men required to sail and maintain the vessel, gun crews were needed, as were marines and many others who were not required in a merchant ship of similar size. Death from battle, accidents and disease, were an accepted fact and, to counteract these losses, every department was grossly over-manned, which of course only compounded the problem... until enough of them had died!

The living accommodations for most of the men were the gundecks – cold and draughty places – and undoubtedly very wet in bad weather. Hammocks were slung on hooks, fourteen inches apart. We thought we were crowded in 1940, when we hung our hammocks on hooks, set at eighteen-inch intervals!

Today a seagoing toilet is known as a 'head' and this term comes from the days of sail, when the only place the crew had to relieve themselves, was at the head of the ship, in the bows of the vessel. That area had a grating, well washed by the sea thrown up by the bows, and this is where the men had to squat to relieve themselves, no doubt getting soaked in the process, if there was anything of a sea running. In really bad weather, it could be a very dangerous place. One wonders what Cook's compassionate and logical mind made of a system that spent time and money training desperately needed men, and then squandered their well-being and their lives in such a profligate fashion.

Whilst *Eagle* and other similar vessels were at the heart of any blockade, they were too big to hazard close inshore. Smaller craft, that were much handier in tight situations, were provided to cover those areas and they were also used further afield, to act as the eyes of the fleet. If need arose, they would stop and search suspect vessels.

On 5th April 1756 James Cook was given command of one of these ships, a forty-foot cutter, not much bigger than the average present-day yacht. These small vessels usually cruised independently for two to three weeks, before returning to the senior ship. Even at this early stage, Cook's log records pilotage details that would be of value to others. This particular cruise lasted just three weeks, in the course of which a Spanish brig, bound for Bilbao, was stopped and searched. Importantly for Cook's future, this cruise showed Palliser that there was a man who could run a ship and maintain good discipline, simply by the exercise of his own power of command. On 25th April Cook took his cutter into Plymouth, where he found the *St Albans* preparing to return to the *Eagle*, so he took passage in her and returned to his ship.

Not long after Cook had rejoined *Eagle*, three French ships were sighted and *Eagle* and *Romney* gave chase. One of the French ships parted from her companions and *Romney* continued in pursuit, finally catching up with her, boarding and capturing her. By noon *Eagle* was close enough to the other French ships to open fire and one of the Frenchmen immediately struck her colours. Captain Palliser put a prize crew aboard the French ship and continued in pursuit of the other, the *Triton*, which eventually surrendered. Cook was given command of the *Triton*, the larger of the two Frenchmen, with orders to take her to London. He handed over his prize in London on 24th of June, complete with its cargo of sugar and coffee. He sent his men to Plymouth and, by the 29th, was following in their wake.

Palliser sailed the *Eagle* directly to Plymouth, where he landed his sick crew members. So many of his men had gone down with scurvy that there were scarcely enough left to work the ship. *Eagle* was restocked with food and stores of all kinds and men were found to replace those who had been put ashore. It was winter now, but the blockade continued, and Eagle took up her station in the Western Approaches. The winter was to be a bad one, with a succession of gales. The inadequate quarters in which the crew were housed must have been constantly wet and cold, their clothing would have never really dried either, making the conditions intolerable. Once fabric is soaked in seawater, even if it is dried, it will continue to attract moisture from the air, until such time as the salt is washed out in fresh water.

In the year 1756, Britain had finally declared war on the French. It was the culmination of a period of tension, which had originated in Europe with the conflict between Prussia and Austria over the control of Silesia (in modern-day Poland), with its valuable coal-mining and textile industries. Frederick II of Prussia had annexed Silesia in 1740, but it was Austria's attempts to regain sovereignty which led to the Seven Years' War. As tension mounted, Europe found itself divided into two camps, with Prussia and Britain on one side and Austria, France, Russia, Sweden and Spain on the other. The conflict was wide-ranging, with battlefields as far apart as Europe, India and North America. Some of the sequels of this war would be to set the scene for the shape of colonialism for the next century and for British world ascendency. It was to be a crucial point in the history of Britain, and an extraordinary figure emerged to take the helm, that of the powerful War Minister, William Pitt the Elder. Pitt had enough vision to realise that if he wanted to hang on to and expand Britain's overseas colonies, he needed to put in place a comprehensive foreign policy, underpinned by a strong military and maritime base. Within a few months, he reorganised the Army and the Navy. George Anson had been placed in charge at the Admiralty, where he was making his presence felt, and, as a result, the English shipyards' order books were filled with work.

One of Pitt's immediate aim was to curb French activities in North America and bring a measure of security to the English colonies there. His strategy involved two thrusts northwards overland and one by sea that would be directed at Quebec.

Back home in the Western Approaches the Navy was doing its best to disrupt France's Atlantic supply ships. This was the background of Cook's early days in the Navy. The blockade was to some extent succeeding in keeping the French bottled up in their harbours, but, by January 1757, *Eagle* could no longer continue with her blockade work and was compelled to return to Plymouth to refit and to replenish her stores. It was May before she was in a good enough state to proceed to sea and, when she did, it was in company with *Medway*. The two ships headed towards their old familiar beat, off Ushant.

At about midnight, on 30th of May, a large French ship was sighted and she was chased throughout the hours of darkness. Come daylight, the quarry was identified as a powerful man of war, the *Duc d'Aquitaine*, with her guns ready for action. Palliser had prepared *Eagle* for battle in the course of the night. Aboard *Medway* no such preparations had been made, until daylight revealed the enemy ship and she lost ground as she shortened sail to clear for action. This left *Eagle* on her own to approach the French vessel. At about 0400 Palliser was about two ships' lengths off the *Duc d'Aquitaine*'s quarter and he opened fire on the enemy. Forty minutes later *Medway* came up on the French ship's stern and added her gunfire to *Eagle*'s. Shortly after the arrival of *Medway*, the *Duc d'Aquitaine* gave up and lowered her ensign in a token of surrender.

The *Duc d'Aquitaine* was a large ship of the French East India Company, of approximately 1,500 tons, armed with fifty eighteen-pounders and a crew of nearly 500 men. She had left Lisbon as a ship of war with the intention of intercepting a British convoy, guarded by a twenty-gun ship. Palliser's successful action saved the convoy which, almost certainly, would have been overwhelmed, had it been engaged by the *Duc d'Aquitaine*. The latter's losses during this engagement were fifty dead, with over a hundred wounded – twenty-two of whom were considered to be serious. *Eagle* had fared rather better with ten men killed and twenty-two wounded, seven dangerously so.

All three masts of the French ship went by the board at the time of her surrender, leaving her helpless in the water. *Eagle* too had suffered serious damage to her rigging, sails and spars and there was much to be done before she could be called seaworthy. A prize crew from the *Medway* looked after the stricken French vessel, while Cook, as *Eagle*'s Bosun, organised the repair and replacement of damaged gear. The weather was deteriorating and the dangerous waters around Ushant were no place for a partially disabled ship to linger. Temporary repairs were made to enable enough sail to be raised and get *Eagle* back to England. There her crew were paid off and she was laid up, so that the damage sustained in her fight with the French ship could be repaired.

As all this was happening, some of Cook's friends back home had written to their Member of Parliament, asking that consideration might be given to obtaining a commission for him. The Secretary for the Navy passed the letter to Captain Palliser. Glad as Palliser would have been to recommend Cook, the latter had only been in the Navy for two-and-a half years and regulations required a minimum of six years service before he could be granted a commission as a Lieutenant. The regulations, however, allowed a Master's Warrant to be granted and Cook received this, along with instructions to join the *Solebay*, a sloop, lying at Leith. One month later Cook was appointed Master of the sixty-four-gun *Pembroke*, under the command of the much-admired and 'truly scientific gentleman', Captain John Simcoe. *Pembroke* was to spend the winter months on blockade duties around the Bay of Biscay, returning to Plymouth in February 1758.

Such was the drive and urgency in Plymouth, that a mere sixteen days later *Pembroke* was on her way again, as part of the strong escort given to the convoy of ships transporting 14,000 of Pitt's best troops to Halifax, Nova Scotia. There, they were to prepare for the attack on Quebec. The strategic importance of Quebec was unsurpassed. The town was protected by high cliffs overlooking the Saint Lawrence River which, at that point, narrowed from several miles wide to a few hundred yards. Vessels could not sail further into Canada, without running the gauntlet of the French guns, positioned on the cliff tops. It was clear that whoever held Quebec, held Canada.

Some of the heavy units of Admiral Edward Boscawen's fleet pressed on ahead, with the intention to stop reinforcements from reaching the French in North America. Unfortunately the weather was against the convoy and, what would normally have

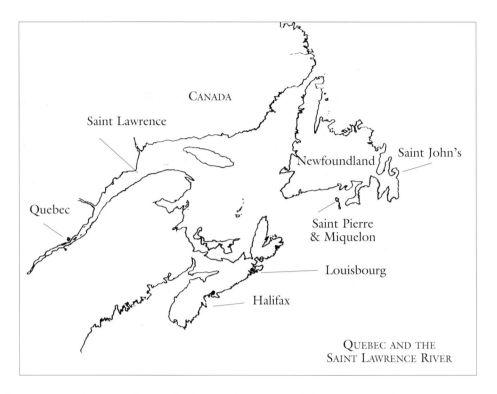

QUEBEC AND THE
SAINT LAWRENCE RIVER

taken two or three weeks, took eleven. So many of the troops and seamen succumbed to scurvy, that little could be achieved that year. The only action worth noting was a land-based attack on the French fort, Duquesne.

General Jeffrey Amherst of the Army, who was to command the troops being convoyed to Halifax, was serving in Europe at the time and was unable to reach Plymouth before the fleet sailed. *Romney* was instructed to wait for him and then to transport him to Halifax.

In the absence of Amherst, a young but brilliant officer, Colonel James Wolfe (he would be made a general after the fall of Louisbourg), was in command at Halifax and he drew up plans for an attack on the strongly defended fortress at Louisbourg. Wolfe's ideas were well in advance of many of his more conventional colleagues, whose idea of a battle plan was a full frontal attack, usually at the best defended point. Wolfe's approach was altogether different. He would keep the enemy occupied with a frontal attack, whilst other troops were stealthily attacking a vulnerable spot, elsewhere in the enemy's defences, and this was the ploy he planned to use at Louisbourg. This strongly fortified settlement had been founded in 1713 and named in honour of the French King, Louis XIV. It stood on an island at the mouth of the Gulf of Saint Lawrence and had an enormous strategic importance in the defence of the whole of the Quebec province. Its economic importance was also paramount for the French, as it was the centre of their fur, fishing and shipbuilding industries.

Amherst arrived, shortly before the attack was due to start. To Amherst, Wolfe's tactics seemed all wrong. He had been brought up in the school that believed that one should concentrate one's strength on one point and go in hard. To divide one's strength, as Wolfe intended to do, was unheard of. Amherst cancelled Wolfe's plan and, true to type, he ordered a full frontal assault against the heavily defended coast, close to Louisbourg.

The attack went in on 8th of June. The landing, led by Colonel Wolfe, was made and held, but at a considerable cost to the troops taking part. There was a heavy surf running, which overturned many of the boats as they arrived under the guns of the well-entrenched defenders. It would be comforting to know that Amherst had a moment of doubt and wondered whether Wolfe's plan would have worked and with fewer casualties! Although the landing was held, no further progress could be made.

Four days later, *Pembroke* arrived from England and berthed at Halifax with large numbers of sick men on board. The soldiers, who were fit enough, disembarked and went to the aid of the men holding the landing area. There followed a month of frustrating stalemate in the course of which Wolfe continually urged that the Navy should be asked to co-operate with the Army. Time was slipping by and the all-too-short summer would soon be coming to an end. Eventually Wolfe prevailed and the Navy was asked to mount an attack on Louisbourg.

Within the harbour was a French squadron of naval vessels. The five largest had been anchored in a line to form a boom across the harbour mouth, thereby denying the British ships access to the harbour. Late in June the navy attacked these vessels, two were destroyed by fire and a third blew up. A few days later a motley collection of fifty open boats was rowed past the remains of the boom into the harbour, under cover of darkness. The intention was to cut the lines of the two remaining French ships and tow them into the north-east basin. They managed it with the *Bien Faisant*, a ship of sixty-four guns, but the other vessel, the seventy-four-gun *Prudent*, was aground and she was destroyed by fire. With the boom removed, the town was open to attack and occupation by the fleet. This left the French Governor no alternative but to surrender on 26th July, following a five-week-long siege. The Saint Lawrence River was now open and the British were primed to wrest Canada from French sovereignty.

Once Louisbourg had fallen, Sir Charles Hardy took a squadron, which included Cook's ship *Pembroke*, to patrol the mouth of the Saint Lawrence River. The Bay of Gaspé, at the southern shore of the Saint Lawrence River's estuary, boasted a number of French settlements. An attack was mounted on them in the course of which a number of buildings and ammunition depots were destroyed.

At the end of September 1758, *Pembroke* and the other ships returned to Halifax, where they were readied for the resumption of the campaign, as soon as the weather improved. Admiral Boscawen and General Wolfe went back to England, as they were both in bad health, leaving Amherst in charge of the remaining British forces. The long, bitterly cold Canadian winter set in, forcing the combatants into inactivity and boredom.

For Cook, these months of relative inactivity turned into a period of exciting study and productive work. The day after the fall of Louisbourg, he had met a young army officer, one of Wolfe's men, Lieutenant Samuel Holland, who was a military engineer and surveyor. He explained to Cook how, by taking angles off fixed points, it was possible to reproduce the features of a landscape on paper. He was at the time preparing a plan of the encampment. Cook, who had already acquired the rudiments of trigonometry and surveying from Captain Palliser, was entranced. The friendship between the two men grew and, during the winter of 1758, Cook and Holland, with the blessings of Captain Simcoe, spent every minute of their free time in the Great Cabin of *Pembroke*, working on charts. Cook had indeed discovered his chief talent.

Below: The Siege of Louisbourg *by Richard Paton*

By another stroke of luck, it took place in one of the great periods of change in naval practices: the rise of 'scientific' navigation.

It was the time when Cook started work on his own 'Sailing Directions'. These navigational aids contained first-hand observations, aimed at other seamen, indicating dangers and giving reference points to enable vessels to sail safely in unknown waters. Cook was not alone in compiling these documents, but his are particularly detailed and painstaking.

With the advancing spring came the renewed urgency to implement Pitt's strategy in North America: the capture of Quebec. Admiral Boscawen was replaced by Admiral Charles Saunders. It was planned to send a squadron to blockade the mouth of the Saint Lawrence River to make sure that French reinforcements could not reach Quebec. *Pembroke* was included in the squadron which was commanded by a Captain Durell. When Saunders arrived in March 1759 to take up his new command, he was angered to find Durell's squadron still at anchor in Halifax. When an explanation was demanded of Durell, he claimed that he did not yet know if the ice had cleared from the Saint Lawrence River. The records say that he was told to go and find out for himself. What a pity that a verbatim record of this conversation is not available! Durell not only learned that the river was clear of ice, but he also discovered that Louis Antoine de Bougainville, at the head of a French squadron, had managed to sail his ships, along with much needed reinforcements and supplies, into the harbour at Quebec. This is the same Louis de Bougainville who would almost beat Cook in the race for the discovery of Australia some ten years later.

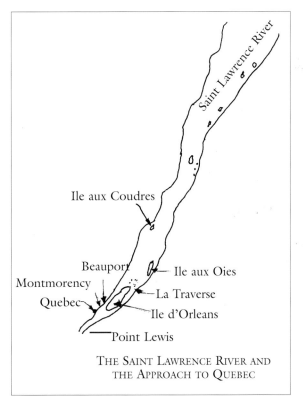

THE SAINT LAWRENCE RIVER AND THE APPROACH TO QUEBEC

Durell redeemed himself somewhat by capturing the Ile aux Coudres, some fifty miles from Quebec which was to serve as a base. The channel leading up to Quebec lies to the north of the island and is reduced to one or two miles in width, making it readily defensible as a base and as a stop to the possibility of traffic going to the relief of Quebec.

In Quebec the French force of 16,000 men was under the command of General Louis Joseph Montcalm, a man of proven ability. The only approach from the sea was by the Saint Lawrence River, which was a very difficult river to navigate, even for small ships. It was beset by fast currents and rocky hazards, a highly dangerous combination for a sailing vessel.

Should the wind drop, the rudder can no longer influence the direction of the ship's head and vessels caught in this situation would be set onto the rocks by the strong currents and wrecked. The French had removed all the buoyage and sea markers in an effort to make things even more dangerous for the potential invaders. Entranched on top of the cliffs, they probably felt quite safe, but they had reckoned without the ability, courage and determination of two thirty-two-year-olds, James Wolfe and James Cook, and the maritime expertise of the Navy.

The normal route for small sailing vessels, as they approached Quebec, was to hold to the northern shore, until about ten miles from their destination, at a point where the river begins to narrow and there are a number of rocky islets ahead, followed by the Ile d'Orleans. From there, the channel crosses the river to the south shore and then passes between the south shore and the Ile d'Orleans. The channel across the river is called the Traverse and is notoriously difficult.

Back in 1711, an English force of 5,000 soldiers, with an escort of twenty or so warships, attempted to reach Quebec by this route. They were defeated by the Traverse. So many ships foundered and so many troops drowned, that the survivors were forced to withdraw.

Knowing this, and since the success of the operation hung on his fleet reaching Quebec, Admiral Saunders ordered that the passage to Quebec be surveyed and charted and he put Cook and Mr Bisset, the Master of Durell's Flag Ship, in charge of the operation. Every night, for a week or more, the Master of each ship provided boats and crews to work under Cook's direction to produce a record of useful shore marks, depths, the behaviour of the currents in the racing river, and noting anything else that might be of value. It was not unusual for the sounding and surveying parties to be attacked by Indians or Frenchmen, but the work went on regardless. It was of such quality, despite these adverse circumstances, that it would take about hundred years before Cook's charts of the Saint Lawrence River were superseded.

On 17th June 1759, exactly a week after the completion of the survey, Saunders and Wolfe, with the troop transports and the war ships, reached Ile aux Coudres. Saunders shifted his flag to the *Stirling Castle*. Cook and his colleagues set about marking the surveyed channel with buoys and, on 24th June, the perilous passage of the Traverse was begun. Within a week, over 200 vessels of the fleet were safely through and anchored at the western end of Ile d'Orleans. The impossible had been achieved.

General Wolfe was back from England and was in charge of the army campaign. Once the island was occupied by British troops, the final preparations for the attack on Quebec could begin. Point Levis, on the south bank of the river, opposite Quebec, was the site chosen on which to erect batteries of thirteen-inch mortars, which were capable of firing a 200 pound projectile some 5,000 yards.

While this work was taking place, Cook, and the Masters from some of the other ships, were busy surveying The Narrows and the river to the west of the city, in preparation for the final assault.

The ships anchored off Ile d'Orleans were often under fire from the shore and, on several occasions, the French made concerted attempts to set fire to the anchored fleet. Six fire ships were used in the first attack – old vessels, filled with tar and pitch and exploding bombs. The guard boats managed to hold the burning ships with grappling hooks and to take them in tow, beaching two on the shore of

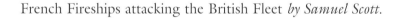

French Fireships attacking the British Fleet *by Samuel Scott.*

Ile d'Orleans, towing the others clear of all the ships, before casting them off to burn themselves down to the waterline and to sink.

A month later, on 28th June, a fire raft at least 200 yards in length moved downstream towards the assembled fleet. It must have been an awesome sight. The blazing raft was made up of a collection of wooden stages, schooners and shallops, chained together and loaded with explosives. The guard boats managed to take it in tow and beach it, before it could do any harm.

General Montcalm was trying to delay the British attack. He may have had the advantage of numbers, but his men were tired and hungry. They had sustained a long siege, first by sea-ice and then by the British fleet, both of which had effectively prevented all reinforcements from reaching the town. He hoped to delay matters till the onset of winter, forcing the invaders to retreat for fear of being caught in the ice.

General Wolfe was well aware of this danger and was anxious to press on. Summer was passing. The British fleet was sailing up and down the river, keeping Montcalm and his men constantly on the alert. Quebec was set on fire and the French scored several hits on British ships.

When Cook's survey and marking of the channel to the north of Ile d'Orleans was completed, the sloop *Porcupine* led the larger ships through to a point where they could anchor, as the first step towards taking Montmorency, to the east of Quebec. Ships had never been through The Narrows before, the Canadian pilots believed this route to be far too hazardous to attempt, even for small vessels. The same night, Wolfe landed his troops and took Montmorency without opposition. Quebec was now threatened on its eastern flank.

The next step in Wolfe's plan required the fleet to move upstream to the west of Quebec, which they did, under the cover of darkness. This manoeuvre was given some protection by the mortar batteries at Point Levis, that laid on a heavy barrage, as the ships negotiated The Narrows. One has to stop and remind oneself that the only motive power available to these large ships in these tortuous channels was the wind. It was a superb display of pilotage and seamanship.

During the night of 12th to 13th September, a diversionary attack was staged by *Pembroke* and other vessels of the fleet, against Beauport, at the eastern side of Quebec. The French moved their reserves to deal with this attack. This gave Wolfe, and a force of 5,000 men, the opportunity to scale the Heights of Abraham, on the western side of Quebec, with a minimum of opposition. This action was followed by a day of bitter fighting and Quebec finally capitulated on 18th September 1759.

Both Wolfe and Montcalm died in the action but, with Quebec in British hands, the campaign was at an end. The fall of Quebec ensured the victory of Britain in the Seven Years' War and The Peace of Paris was signed a few months later. Once the troops were in winter quarters, the fleet dispersed, most of the ships returning to England to refit.

Captain Simcoe, who had been such a good mentor for Cook, had succumbed to illness on 15th May. On 23rd September 1759 Cook was transferred to the larger vessel, *Northumberland*, under the command of Admiral Lord Colville, who was also to become Cook's lifelong friend. *Northumberland* and six other ships, were left behind, with Lord Colville as Commander in Chief of the North American Station. Cook carried on with his unofficial survey work and his sailing directions.

In early 1761 Lord Colville, in recompense for Cook's 'indefatigable industry in making himself Master of the pilotage of the River of Saint Lawrence,' granted him a bonus of £50 (Cook's normal pay was six guineas a month). Lord Colville was also to address the Admiralty, referring to 'Mr Cook's genius and capacity.' His and other well-wishers' recommendations were to pave the way for James Cook's future career. On 7th October 1762, *Northumberland*, with Cook aboard, finally left Canada, headed for home.

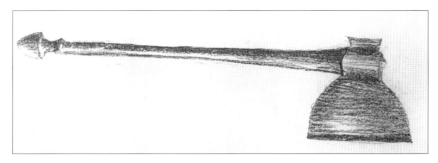

Indian axe

4. Surveyor to the King

November 1762 to November 1767

By the time Quebec was captured, James Cook had been in the Navy for just over four years. For two of those years he had served as a Master, an achievement in itself, and one which would have satisfied many lower-deck men. Cook did not stop there: he was quickly recognised as one of the most skilful Masters in the Navy and tipped for advancement by men of influence and authority.

James Cook's rapid rise must have been due partly to his being brought to the notice of the right people at the right time, but, without his outstanding qualities, those same people would never have encouraged and nurtured his career as they did. There is no doubt that Cook was an ambitious man, but his ambition manifested itself in his desire to excel at what he was doing, rather than in a mere race to climb the ladder to fame. Indeed, his modesty, allied with his other qualities, is what makes the man such an appealing character. He was one of those fortunate people who love the work in which they are involved. He had the kind of mind that *must* know, *must* understand; he was always seeking further knowledge in his chosen field, and, to satisfy him, that knowledge had to be put into practice.

Northumberland returned to England and, by 11th November 1762, Cook left his ship for a short spell ashore. Following a two-week courtship, he married Elizabeth Batts on 22nd December. At twenty-one, she was thirteen years his junior and, although her family and friends raised doubts about the marriage, it was to be an enduring and passionate relationship. Despite their long separations, caused by James's long periods at sea, he remained faithful to her, resisting the temptations of the sirens during his journeys. She would outlive him by fifty-six years and never remarried, burning the letters Cook had sent her from his long years at sea, taking with her to the grave the secret of their long love affair.

They were poles apart and meeting and marrying her evoked in Cook a sense of wonder which would never pale. His background was humble, to say the least. She belonged to the merchant's class, was light-hearted and romantic, while he was aloof and scientific. She enjoyed being the centre of attention and would relish rubbing with all manner of grand people, as her husband's career took off and his fame grew. From all accounts, she worshipped him and encouraged him in his dreams. They were soul mates. The going away to sea must have been agony and the reunions bliss.

They first set up home in Shadwell, a Thames-side village, a mile or two downstream from the Tower of London. They were left in peace until early March 1763, when Cook received a summons from the Admiralty.

Commodore Thomas Graves, who was the Governor of Newfoundland, had been pressing the Admiralty to allow Cook to complete the survey of Newfoundland. On 19th April Cook eventually received orders to join *Antelope*, which was to carry Graves and his staff to Newfoundland, where they arrived in June 1763.

Cook's first task was to survey the islets of St Pierre and Miquelon. The Treaty of Paris had granted these two tiny islands, that lie just off the south coast of Newfoundland, to France. With the islands went fishing rights, which were of some importance to the French. More important to the Admiralty was the possible use by France of the islets as bases in any future conflict. Obviously, a knowledge of the topography of the islands would be useful information, should such an eventuality arise, hence the urgency of Cook's survey.

Long before Cook had completed his work, the French representative, M d'Anjac, arrived aboard a frigate of the French navy, accompanied by some transports, with the intention of establishing a settlement, as soon as the British had handed over the islands. Cook needed time to complete his survey and the Governor of Newfoundland and his staff pulled every bureaucratic gambit out of the bag to delay the hand-over, much to the fury of M d'Anjac. This diplomatic procrastination was successful, however, and enabled Cook to finish his task, before the French took possession of the islets.

Cook was now free to make a start on the survey of Newfoundland. He worked on the southern coast, until winter set in and made it impossible for him to continue. He was able to leave for England and his family – the first of his six children, James – had been born six weeks earlier. Cook used his war payoff to buy a brick house, situated at number 6, Mile End Road in the East End of London. This modest house was to be Cook's home until his death. Elizabeth herself, was to continue living there for several years after his death. But his first leave as a married man, all too soon came to an end on 23rd April 1764.

During the survey work carried out by Cook in the course of the previous summer, he had been transported aboard *Antelope* or *Tweed*, as was necessary, but this was not an entirely satisfactory arrangement. Graves had been replaced by Palliser as Governor of Newfoundland and Palliser pressed the Admiralty to give Cook command of a small vessel, that could be fitted out as a survey ship. A schooner named *Grenville* and lying at St Johns was deemed suitable for the job and James Cook was given command of her under the overall command of Commodore Palliser. His pay remained the same, but the title of 'King's Surveyor' was appended to his name and he arrived at St Johns on 13th June 1764 to take up his post the next day and to start the season's survey work. This time he made a start with the northernmost part of Newfoundland and its extreme northwest coast.

Some time late in the season, Cook had a most unfortunate accident. The powder flask he was holding, exploded and caused considerable damage to his hand, also injuring a man standing close by. There was no surgeon aboard *Grenville*, but one was believed to be aboard a French fishing vessel, lying in Noddy Harbour, and they made for it. Cook's right hand was almost torn off, the thumb hanging by a tendon. Their information proved to be right and the two injured men were treated by the French surgeon. Cook was in fact very lucky. Had a similar injury been treated in one of the London hospitals, where disease and filth were rife, he would have probably lost, not only the hand, but the arm as well. In the unpolluted air of Newfoundland, the skilful surgeon saved the hand and Cook was back surveying within a month. The mutilation was so severe, however, that it was to be the means of identifying Cook's remains, fifteen years later.

On 1st October of the same year, one of *Grenville's* boats that was out sounding, struck a rocky ledge, knocked a hole in its bilge and filled quite quickly. Fortunately the cutter was on hand and was able to rescue the men from the sinking boat. It is usually claimed that Cook was one of those thrown into the water and that he was lucky to escape with his life. There is no mention of this incident in Cook's log, but knowing how Cook consistently failed to record mishaps to his person, it is neither evidence for or against the story.

By now the season was drawing to a close and *Grenville* returned to St Johns to re-provision and refurbish for the journey across the North Atlantic to England. On 12th December 1764 she was berthed at Woolwich and Cook was on his way home to his family. When he arrived, he found his mother-in-law in charge, the midwife in attendance and Elizabeth about to produce their second son who was to be named Nathaniel. Cook's first born was exactly one-year-old.

Grenville was being completely re-rigged and, naturally, Cook kept a close watch on the work. He obviously commuted to *Grenville* on a regular basis as the daily log is

Opposite: Portrait of the Earl of Sandwich *by Thomas Gainsborough*

maintained until 15th January 1765, after which he retires to his family for some ten weeks. For the next three years, life for Cook was to continue in much the same manner, eight months away and four months at home.

In the course of the 1766 season, in addition to continuing with the survey of Newfoundland, Cook found time to observe the eclipse of the sun. The data he collected enabled Dr John Bevis of the Royal Society to calculate the longitude of Newfoundland. The calculation of longitude was one of the big scientific problems of the time; more about this later. This set of observations brought Cook to the notice of the Royal Society and raised his status in their eyes, from a 'mere ship's Master and chart maker,' to that of a learned man of science, a reputation that was to serve him well in days to come.

Cook's final season in Newfoundland was spent surveying and charting, adding to the work he had already done. It had gone well, winter was approaching and it was time to leave. By early November Cook brought *Grenville* back to England and was close to the Thames Estuary, off Deal. On 9th November, a pilot was picked up to help them thread their way through the difficult waters ahead. *Grenville* anchored that afternoon, just above the Nore Lighthouse, in strong winds that harboured some nasty squalls. The Nore Sand, which the light guarded, is just a couple of miles north of Sheerness.

We are so accustomed to the Thames and other waters around Britain being marked with reliable light buoys these days that we perhaps lose sight of the fact that in Cook's day there were buoys but very few lights. What lights there were, would be provided by a fire burning in a brazier on top of a pylon, built in shoal water. Some unfortunate character would be rowed out each evening, summer and winter, with a bag of coal and some meagre rations, to tend the fire through the night. These pylons were, for obvious reasons, called 'beacons'. Many of them still exist to this day and provide useful day marks for small boats, but they no longer have a light. In spite of this, they are still called beacons.

With so few lights to show the way, sailing at night through an intricate area such as the Thames Estuary, would be an act of folly, especially if the sky was overcast, as the report of strong winds suggests. At four pm on a November afternoon, it would certainly have been dark already and I suspect that a prudent master would have anchored for the night, well before that.

The bad weather continued throughout the night and, at some point, one of the anchor cables parted, allowing the vessel to swing round, so that she struck the sand very heavily. Not long after, the other anchor cable also parted and *Grenville* went adrift, fetching up on Knock Sand, about three miles to the north. She lay quietly, as the tide ebbed and left her high and dry, but some hours later, when the rising tide brought back wind-whipped waves, the stricken boat was repeatedly lifted and slammed down hard onto Knock Sand. This hazard in the Thames Estuary is composed of compacted sand and can make a hole in the bottom of a vessel in very little time, if the seas are of any size.

It was obvious that the gale would be with them for a while and so preparations were made to abandon ship, while it was still possible. The decks were cleared of loose gear and all the hatches were secured. The ship's boats were hoisted out and by midnight the crew were on their way to Sheerness.

The next morning the weather was much improved, the crew lightened *Grenville* further and she came off the sand without difficulty. By 15th November 1767 Cook berthed his ship at Woolwich and, undoubtedly, made his way post haste to Mile End Road and his family.

Over the past years, James Cook had made his mark in many ways. His superiors knew him as a man gifted with that all-important, but intangible quality the navy calls 'power of command', a presence that inspires respect and obedience among subordinates. He was known as a man, who could work without supervision, with all the talent and ability required of a first-class seaman. His scientific approach had also attracted the attention of the Royal Society. The stage was set for the next act.

A calm

5. Preparations for Cook's First Voyage

May to July 1768

As mentioned on page 29, in 1679, Sir Edmund Halley – of Halley's Comet fame – read a paper before the Royal Society. He explained that in 1762 and again in 1769, the planet Venus would pass between the sun and the earth and would be visible against the disc of the sun. He went on to say that if observations of this event were made at widely spaced stations around the world, astronomers would, for the first time, be able to calculate the distance between the sun and the earth with some accuracy.

Governments were awakening to the value of scientific discovery and the prestige that came with it. As a result, a number of countries were prepared to help finance twenty astronomers from Britain and other European countries, who were to observe the 1762 Transit of Venus. Accordingly, the astronomers were sent to various places around the world to observe the phenomenon, but the results proved disappointing.

The chosen positions were not sufficiently far apart to make the results meaningful and the instruments used by many of the observers lacked the precision required for the task. And then, of course, there was the weather!

One observer, a Frenchman named Le Gentil, intended to sail to India to observe the Transit of Venus, but he was badly delayed by foul weather and was still at sea on the appointed day. He attempted to make his observations from the deck of the ship, but, not surprisingly, he found it impossible to work with the

A Gregorian reflecting telescope, similar to the ones used for the Transit of Venus. The Royal Society supplied the astronomers with two and Cook brought a third one which he had used in Newfoundland, aboard Grenville.

necessary accuracy. Le Gentil decided to remain in India and observe the next transit, which would take place in seven years time. He rose early on the day, only to find the sky completely overcast. The next transit was not due for over a hundred years, so he went back to France, only to discover that his family members had presumed him dead and had already divided up his estate among them!

The Royal Society was determined to avoid the disappointments of the 1762 expedition. To ensure this, it was essential to have an observer in the South Pacific and King George III was asked to help finance the project. The King was sympathetic and instructed the Navy to provide a ship. The King's request for a vessel to take the scientific party to the South Pacific coincided with the Admiralty's intention to send a ship to the South Seas to explore and annex any land discovered.

The old enemies, the French and the Spaniards, were not entirely happy to have Britain trampling over an area they considered to be their domain, but they reckoned that a scientific expedition would advance knowledge, from which all would profit, so they raised no objections. The government was happy to oblige the Royal Society and, when the Transit of Venus had been observed, the real business of the expedition could be conducted, that is to say, exploration followed by exploitation.

The Royal Society chose Alexander Dalrymple for the post of principal astronomer, whose role would be to observe the Transit in the South Pacific. He was invited to meet the Council to discuss arrangements for the voyage. In the course of these consultations he made it clear that he would not be satisfied with being a passenger, but would expect to have command of the ship in which he was to travel.

Some time in March the Royal Society received a letter from the Admiralty in which they were told that Admiralty surveyors were looking for a suitable craft and that when a likely vessel had been found Mr Dalrymple would be consulted. On April the first, the Admiralty wrote again to announce that a vessel had been purchased and would the Royal Society tell them how many passengers there would be, and an indication of the 'quality' of those passengers. They also wished to know if there were any special instructions the commander of the vessel would need to be given.

Clearly the Admiralty did not intend for Dalrymple to be the commander of the ship. For the latter, it must have been a devastating blow. He was not only an astronomer, but he also had some knowledge of navigation and was something of an authority on the Southern Hemisphere. He had argued passionately for the existence of an undiscovered Southern Continent, or *Terra Australis Incognita*. A vain, self-educated man, he was convinced that no one was better qualified than he, to lead the Transit expedition. As we have seen, he had already mentioned publicly that he would not consider sailing as a mere passenger, but expected to have command of the vessel supplied by the Admiralty. He saw the Navy's rejection as a great personal slight.

Lord Morton, the President of the Royal Society, went to see Sir Edward Hawke, the First Lord of the Admiralty, hoping for a reversal of their decision. But the First Lord made it abundantly clear that there could be no possibility of a non-Navy man taking charge of one of His Majesty's ships, swearing that he would 'rather his right hand be cut off than see a civilian command a naval vessel.'

The President of the Royal Society reported what had transpired back to the Council. Dalrymple was present and he said again that he would not travel, unless he was in charge of the vessel. This left the Council with no choice, but to look for another observer.

Captain John Campbell was a member of the Royal Society and a friend of Edward Hawke, the First Lord. After discussing the matter with him, Campbell took it upon himself to introduce Cook to the Society. Cook met the Council and accepted their offer to observe the Transit, along with Charles Green, the assistant to the Astronomer Royal. Palliser, who had been Cook's commander for so long, was delighted to recommend him unreservedly. On 25th May 1768 Cook was promoted from Master to the commissioned rank of Lieutenant and given command of the vessel.

The ship that had been purchased for this project was a Whitby collier of the kind on which Cook had cut his nautical teeth and named the *Earl of Pembroke*. The Navy renamed her *Endeavour*.

It seems that Cook had no hand in the choice of the vessel, indeed when it was purchased he was working aboard his ship, *Grenville*, getting her ready for her next assignment. But, had he been consulted, he might well have chosen *Endeavour*, or something like her, himself.

The Earl of Pembroke leaving Whitby *by Thomas Luny. This collier was later renamed* Endeavour *and put under James Cook's command.*

A collier might seem to be a strange craft to choose for the exploration of unknown and distant waters, yet this type of vessels might have been made for the job. Together with the great carrying capacity of colliers, went a comparatively shallow draught which enabled them to nose their way inshore safely, along an uncharted coast. Colliers were designed to make their way into rivers like the Thames so they had to be manoeuvrable. Theirs was a fairly flat bottom that would allow them to sit, more or less upright, should they take the ground. Also, the hull shape was such as to make it possible to careen the vessel, should the underwater area need to be repaired or cleaned of weeds and barnacles. It might be thought that a vessel that had been designed to sail in the North Sea and the Thames Estuary would not be suitable for the open ocean, but the North Sea can be formidable waters on which to sail. If a vessel can cope with all that it can throw at it, it can sail anywhere.

Endeavour's vital statistics were something like 110 feet in overall length, with a beam of a little over twenty-nine feet. Her draught was about fourteen feet and she was rated at 368 tons. A ton in this context is not a measure of weight, but of volume. Wine was transported in barrels of a standard size called tuns and the number of tuns a vessel could carry became a way of indicating its size. Later, the spelling was altered to 'tons'.

Endeavour had three masts: the main mast, so-called because it was the tallest, was

placed midships, the fore mast was a little shorter and the mizzen mast (the after-most mast) was shorter still. Such a configuration puts *Endeavour* into the realm of barques or barquentines. Over the years, the definition of these has changed a little, but 'barque' is probably the most appropriate name for *Endeavour* and the one we shall use. Just to add a dash of confusion, the Navy of the day referred to vessels of this kind and size as sloops, but today a sloop is a single-masted sailing vessel!

A major problem for wooden boats was, and still is, the need to protect the underwater area from the depredations of a wide variety of worms whose favourite food happens to be timber. Tropical waters have the grandaddy of them all, the teredo. This worm tunnels into the timber of a wooden boat, coating the three-quarter inch diameter tunnel with a thin film of silica-like material, as it happily chews its way forward.

In Cook's day, as had been done for many years before, British ships were given an extra thickness of planking, nailed to the outside of the hull as a sacrificial layer. A coating of tar and oakum was applied to the hull of the ship, before the sacrificial layer of timber was nailed in place. Some time before Captain Cook was to set sail on his first voyage, the Admiralty had also experimented with sheathing the hull of a number of vessels with sheets of copper as a protection against fouling and it had proved to be successful.

Cook wanted to use copper sheathing to protect the hull of *Endeavour*, but he was concerned that it could be damaged, or come adrift, and that there would be no dockyard available to replace it. The solution was brilliantly simple. The usual coating of oakum and tar was applied to the hull and the sacrificial timber was nailed over it, using thousands of copper nails placed so closely that their large, flat heads touched each other.

There was much to be done to convert the *Earl of Pembroke*, a collier, into HMS *Endeavour*, a survey ship. An early task was to make the *Endeavour* suitable to house the ninety-five people who were to travel in her. Cabins were built at the after end for the officers, the party of scientists and their assistants from the Royal Society. The great open space that was the hold, had been designed to accept a considerable load of coal. The hold was so deep that it was possible to build a second deck for the length of the barque. Part of this extra space was fitted out as accommodation for the crew. Parallel pairs of beams were built in, about eight feet apart and stout iron hooks fixed to them, at fourteen-inch intervals. Each man was allocated a pair of hooks from which to sling his hammock. Tables and benches completed the crew's quarters. A galley was constructed and a working space for the ship's surgeon was also built into this area, complete with a store for his equipment and medicines. Stores of every kind: timber for repairs, tools, blocks, rope and cordage, sail cloth, and the thousand-and-one other items needed to keep a vessel; in a seaworthy condition for at least two years. Powder and shot had to be stored safely, small arms and ammunition; food, water and wine all needed to be stowed somewhere. It was as well *Endeavour* had been designed as a load carrier, because, in addition to the ninety-five people aboard, there was the gear required by the artists, astronomers, two naturalists and the equipment supplied by the Royal Society – all took up a great deal of space.

In addition to this, the Admiralty had given Cook equipment they wanted tested. With her hull shape, the great weight of stores and, indeed, the considerable number of people she was to carry, *Endeavour* would never win races, but she proved ideal for the tasks that lay ahead of her.

Provision had to be made for twelve carriage guns and ten swivel guns which would be installed later. The area where the cargo hatches had been, was now decked over and companionways provided, wherever necessary, to give access to the lower decks. Larger and more powerful bilge pumps were fitted and several tons of iron ballast was stowed in the bilge. The masts and yards were taken out of her, examined and repaired, the rigging was overhauled and replaced, and new sails were provided. Three large open boats had to be stowed on deck and lifting gear installed to enable them to be launched. It seems that the Admiralty met virtually every one of Cook's requests without question.

Cook had been given a good crew, five of the men had moved with him from the *Grenville*, one of them being his personal servant. It is obvious that these five were men who admired and respected Cook and were even devoted to him, although he could be curt, if he thought that a man was failing to perform to his own high standards. The respect for Cook, shown by those who had served with him before, helped the new crew members to settle in with the minimum of friction.

The party of supernumeraries included botanical artists, draughtsmen, naturalists, astronomers and a doctor, some with skills in more than one discipline. The best known among the group were Sir Joseph Banks, Dr Daniel Carl Solander, Herman Spöring, Charles Green, Sydney Parkinson and Alexander Buchan, all men of competence in their chosen fields.

The expedition, headed by a man like Cook, whose wide-ranging abilities were combined with unique organisational skills, had a better chance of running smoothly than any mounted before. Nevertheless, when all the scientific gear, the servants, the assistants and baggage were loaded on board, and in spite of Cook's all-embracing preparations, conditions aboard *Endeavour* were very cramped.

Cook's instructions were to proceed to Plymouth where the crew would be given an advance of two months' pay. From Plymouth he was to go to Madeira and take on board as much wine as he could stow. He was required to enter the Pacific via Cape Horn. At his discretion, he was permitted to land on the coast of Brazil, or the Falkland Islands, if he needed fresh food or water. Once into the Pacific he was to proceed to one of the islands of the Marquesas, where he and Green were to observe and record the Transit of the planet Venus. If he was unable to reach this location in time, he was to find a suitable situation elsewhere, but it had to be in the list of latitudes and longitudes provided by the Royal Society.

Once the Transit had been observed, Cook was instructed to head south and conduct a search for *Terra Australis Incognita*. He was required to search for this continent as far south as 40º of latitude and as far west as New Zealand. If the territory was found, he was to lay claim to it, map its coastline and offshore hazards, note the bays and islands and so on. The Royal Society wanted to know what kinds of animals, trees, fruits and grains were to be found there; specimens were to be brought home. If native people were found, Cook was instructed to befriend them. Failing to discover the elusive continent, he was to continue westwards, until he fell in with the coast of New Zealand. He was to establish its position and explore as much of its coastline as was practicable. Any other islands that he should encounter in the course of carrying out these instructions, were to be claimed for England and their positions noted.

The Navy had already sent Byron and then Wallis to the Pacific with limited success. However, Wallis had returned, in time for Cook to learn that he had found Tahiti and fixed its position with some certainty. This would make it easy for Cook to find the island again. Wallis was also able to report that, after a difficult start, he had gained the trust of the Tahitians and that there were plentiful supplies of fresh food available on the island.

It was realised that the Transit would be visible from Tahiti and so Cook's instructions were amended to allow him to use Tahiti as a base, instead of searching for the Marquesas, which had last been sighted by Mendaña, about hundred years before Cook's departure, but whose position was somewhat vague.

6. England to Tahiti

July 1768 to July 1769

HMS *Endeavour* left Deptford basin on 30th July 1768 and made her way down to Galleons Reach, where she anchored to take in her guns, powder and shot. When this was completed and the anchor was brought aboard and stowed, she made her way downstream. Once *Endeavour* had rounded the North Foreland, she was in the Channel and on her way, reaching Plymouth on 13th August.

Most of the scientific party were waiting to embark and, as soon as they were aboard, together with their mountains of baggage, they discovered that in spite of all the alterations that had been made to the vessel, there was still a shortage of accommodation. Shipwrights were brought aboard and they worked for five days, constructing more cabins, while a steady stream of provisions was still being loaded. Joseph Banks and Daniel Solander were the last to arrive. Once they were settled in, the party was complete.

By 26th August *Endeavour* was as ready as a ship ever would be and she took her departure towards Madeira, with a fair wind to help her on her way. They were fortunate with the weather as far as Madeira, where they anchored off Funchal. Unfortunately, as they were anchoring, the Master's Mate became entangled with the anchor cable and was taken down with it. By the time they brought him back to the surface, he was dead.

At Funchal they purchased some 3,000 gallons of wine, enough onions to give each man thirty pounds and a great deal of fresh fruit. The water casks were filled and a 600-pound steer was brought on board. The beast was slaughtered before they left and served as fresh beef on subsequent days, instead of the ubiquitous salt beef.

During the stay at Funchal, a seaman and a marine complained about the quality of the 'fresh' beef they were served, probably with some reason, but they were unwise enough to be insubordinate about it. This earned them twelve lashes each, which were administered with the usual pomp and ceremony. Cook was a humanitarian and did not run lightly to severe punishments; at the time a similar offence would have been likely to earn the culprits a much harsher sentence from other captains. When judging this incident, it must be remembered that eighteenth-century attitudes to corporal punishment, both afloat and ashore, were, quite different from ours.

There is another factor that probably influenced Cook on this occasion. He was in the process of introducing changes in the diet of his ship's company, in the hope that the incidence of scurvy could be drastically lowered, or better still, that this scourge of long sea voyages could be defeated. Clearly, this was no time to allow members of the crew to be picky about the 'fresh' food they were given.

Cook was familiar with Dr Lind's findings on the causes of scurvy and was determined that his ship's company would not be decimated by this avoidable disease. He studied Lind's theory seriously and provisioned his ship accordingly. Throughout his voyages, he never missed a chance to collect or purchase anti-scorbutic food.

Cook's problem was not so much what to take on board, but how to persuade the lower deck to eat it. Much as the crew might deplore the salt beef and hard tack to which they had become accustomed, try changing their diet and there was bound to be trouble. Sauerkraut, for instance, was deemed to be of value in the fight against scurvy and a great many barrels of it were taken aboard. Cook's approach was to leave the barrel of sauerkraut in the galley so that each man could take as much, or as little as he wished. The sauerkraut for the officers' table was set out on dishes which were carried aft with the rest of their food and with as much ostentation as possible. Within a week, Cook had to start rationing the sauerkraut! It has since been shown that sauerkraut was of limited value in the fight against scurvy, but it played its part in breaking down the crewmen's prejudices against unusual foodstuff. When *Endeavour* left Madeira, fishhooks and lines were issued to the crew and, from then on, their diet included fresh fish.

Cook was insistent that the crew's quarters were kept spotlessly clean and he also deplored the fetid atmosphere that so often prevailed in the lower decks, naval architects having given no thought to ventilation. To overcome this, Cook had coal or charcoal fires set in iron pots and placed at various strategic positions, such as companionways, at least twice a week. As the fires heated, the stale air rose and escaped, drawing in fresh air in the process. With these measures, and Cook's

Opposite. *Joseph Banks painted by Sir Joshua Reynolds. Aged twenty-five, Banks was already a respected naturalist and a member of the prestigious Royal Society. Handsome, wealthy and erudite, he contributed no less than £10,000 towards the cost of Cook's first voyage. This, coupled with his influence with the other Fellows at the Royal Society, enabled him to obtain a place aboard* Endeavour, *as the most distinguished of the 'supernumeraries', the non-naval scientists who took part in the expedition.*

insistence that anti-scorbutic foods must be part of everyone's diet, not a single man on board was sick at any time after leaving England. This happy state of affairs was to continue until *Endeavour* was halfway across the Pacific. Even if that had been Cook's only achievement, I suspect that his name would be remembered for it.

From Madeira, *Endeavour* headed south for Brazil. As soon as they had crossed the equator, Father Neptune was invited aboard and, in the time-honoured tradition, those who could not prove that they had crossed the line before, were given the choice of forfeiting four days' wine ration or being ducked in the sea, presumably not before the usual indignities had been performed upon them. It seems that very few opted to buy their way out of the traditional ducking.

With the ceremony completed, *Endeavour* resumed her course for Rio de Janeiro where Cook hoped to replenish his stores, there being little hope of another chance until they were well into the Pacific. Some time before this, the Pope had divided the world into two halves and had kindly given one to Spain and Portugal the other, Brazil finding itself in the Portuguese half. At the time of Cook's visit to Rio, it was firmly in Portuguese hands and only Portuguese vessels were permitted to trade there, while other nationalities could purchase bare necessities only.

On 14th November *Endeavour* hove to outside Rio Harbour and First Lieutenant Hicks was taken in the pinnace to inform the authorities of their identity and their reason for calling at Rio. Hicks failed to return and Cook took his ship into the harbour and anchored. That evening, the pinnace returned with a Portuguese officer on board, who informed Cook that Hicks was being detained, until Cook himself went ashore. He was followed by a guard boat, full of armed soldiers. Yet another boat arrived, this one with a party of Portuguese officers, who wanted to know the purpose of the ship's visit. Satisfied with Cook's reply, they returned to the shore and Hicks was finally allowed to rejoin *Endeavour*.

The next morning Cook went ashore to visit the Viceroy who gave him permission to purchase whatever provisions he required and to take fresh water from the fountain in the centre of the town, but Cook's request for Banks and others to be allowed to go ashore to collect botanical specimens was refused. The Viceroy was deeply suspicious and was incapable of understanding or believing in the scientific nature of the voyage. It seems that the meeting was a prickly one. On one hand, we have an unyielding Cook, justly proud of his position as a representative of His Majesty's Navy, who is not believed when he explains the reason for his visit. He is escorted every time he steps ashore and his scientific party is refused permission to go and collect botanical specimens. On the other hand, we have the Portuguese

Viceroy, very conscious of the English reputation for piracy, confronted with the Captain of this ship which is clearly a merchant vessel, whilst he claims it to be a ship of the Royal Navy. He has a Commission, it is true, but Commissions can be forged. As for this cock-and-bull story about going all the way to the Pacific to look at a planet, or the request to be allowed to put ashore, what are obviously spies, on the pretext that they want to pick flowers... I don't know what the Portuguese version is for 'pull the other one, it's got bells on it,' but I'm willing to bet it got an airing, as soon as Cook had left. During the remainder of their stay, Cook went ashore no more than was absolutely necessary but, typically, he did not waste his time: surreptitiously, he made a chart of the harbour and detailed plans of the fortifications.

There were a few other minor irritations before they were ready to leave but, finally, the day did come. It was up anchor and away, only to have two shots fired at the ship from the fort, as they approached the entrance to the harbour. *Endeavour* came to, anchored and waited. Apparently, the Viceroy's permit that allowed Cook to leave had not reached the fort, even though the date of departure had been known for at least a week. It was not until later in the day that *Endeavour* was finally allowed to sail.

As soon as *Endeavour* was clear of the land, a course was laid that would take them towards the Horn. It took them over a month before land was sighted and, when they did, it was the eastern coast of Tierra del Fuego and the off-lying Staten Island. The passage between the two islands is called Le Maire Straits.

In those days, the recommended route for seamen was to leave Staten Island to the west, make progress southwards in the open sea to latitude 60° South, and then head westwards. Cook chose to go through the Le Maire Straits instead. What prompted him to go that way? It seems to me that there can only be two answers. First, and perhaps foremost, his insatiable curiosity demanded that he see Tierra del Fuego, Cape Horn and the surrounding land for himself; secondly, he perhaps could see no real logic in going so far east and south, when he wanted to go west. He was fortunate that the weather appears to have been kinder to him, than it had been at times to others. Had it blown up rough for him, perhaps he would have taken a different course. Having said that, if the weather *had* been bad, chances are that the wind would have come from the west. The course Cook took would have put him under the lee of the land and given *Endeavour* some protection from the heavy seas that bad weather generates in this area. Cook was an innovator and he chose the route that would become standard among the clipper ships of years to come.

While *Endeavour* stood off, a party was sent to find a bay in which to anchor, with a view to replenishing the ship's supply of wood, fresh water and green food stuffs. Banks and Solander went ashore with this party and came back with nearly a hundred totally unknown species of plants. Molyneux, the Master, found what is now called the Bay of Good Success and *Endeavour* made her way in and anchored within its shelter. About thirty of the local inhabitants were waiting for the party from the ship to come ashore. Considering the climate, they were scantily clothed and their condition was described by Cook as wretched.

The next day the scientific party of twelve went ashore to explore, expecting to be back on board before nightfall. The plan was to climb up through the forest on to the plain they could see from the shore and, from there, to the ridge beyond. The party of scientists and their companions had more enthusiasm for than experience of trekking across unknown territory and they soon got themselves into serious trouble.

The climb through the forest was difficult and tiring and, when they did break through to what they had assumed was a plain and likely to be easy-going, they found it was swampy and covered with tree stumps and three-foot-high scrub. By now it was well into the afternoon and they really should have turned back, but they pressed on. Progress through the thick scrub was painfully slow but they were still intent on reaching the ridge. This difficult area was some two miles across and when they were about halfway, Alexander Buchan, the botanical artist, had an epileptic fit. They lit a fire to make him more comfortable and, four of them, Solander, Banks, Monkhouse, Green and two servants continued their climb, leaving the rest of the party with Buchan.

As the sun set, it began to get colder and a light sprinkling of snow did little to help. They were now in real danger, but did not seem to realise it. When the small group that had gone on ahead returned to the main party, they had a further mile of difficult terrain to cover, before they reached the relative shelter of the trees.

The darkness of an overcast night was upon them, it was getting colder all the time, and the snow was no longer a light sprinkling, but falling heavily. To crown it all, they had no real idea of where they were, but they still didn't seem to realise the seriousness of their situation. Solander and the two servants were exhausted. The servants laid down on the snow-covered ground and were soon fast asleep, a sleep from which they could not be roused. Solander was little better than the two servants and he was half-dragged, half-walked the quarter of a mile to the rough shelter and the fire that had been built among the trees. Men were sent back to bring the servants, but when they found them, they could not wake them and they were too heavy to be carried over the rough, boggy ground to the shelter. They covered them with vegetation, hoping to keep the worst of the cold from them, but the next morning, both men were found dead. Warmed by the fire, Buchan and Solander had made a good recovery from their ordeal. As the sun began its daily climb, the temperature rose with it and the party set off, under the impression that they had at least a day's march ahead of them. Much to their surprise, they found the beach and the sea in less than three hours.

One can imagine Cook's concern and frustration. It would have been foolhardy to send a rescue party while it was still dark. It snowed quite heavily during much of the night and he had no idea what direction the party had taken. He could only hope that Banks would learn to temper his enthusiasm with a little prudence on future exploratory trips!

Endeavour left her anchorage in moderate weather and worked her way south and west towards Cape Horn, all the time taking soundings, making observations and noting the behaviour of the winds and currents. The weather was reasonable but well interspersed with a series of gales and squalls as *Endeavour* made her way towards the Horn. Off the Horn the weather was kind to Cook, to such an extent, that he was able to heave to and take a series of observations to ascertain its position.

Captain Cook established the latitude of Cape Horn with an error of only one mile. His lunar distance sights gave him a longitude that was 40 nautical miles in error. This illustrates the lack of accuracy, inherent to the lunar distance method, when used on board ship. Here we have a distinguished navigator and the Assistant to the Astronomer Royal, in gentle weather, and their calculations of the longitude is forty miles adrift! How well would other ships' Masters, who did not have the combined mathematical expertise of Captain Cook and Charles Green, have fared? I wonder.

The winds Cook experienced when he rounded the Horn were so gentle that he had to set his studding sails (usually pronounced 'stun's'ls'). These are narrow, lightweight, supplementary sails that are rigged on small spars, on either side of the usual sails. They increase the sail area and make the most of the available wind.

Having passed the Horn, Cook headed south and west to search for *Terra Australis Incognita*. He went as far south as 60º, and about 600 miles westwards, and found nothing. We now know that there is nothing to impede the progress of the big seas that can build up in those latitudes, all the way round the globe. Their unrestricted movement gives these seas a quality that is missing from seas that have not travelled such vast distances, and a seaman as sensitive as Cook would have understood the significance of these observations. Nevertheless, his orders were to prove or disprove the existence of a southern continent, and that is what he intended to do.

The search for *Terra Australis* had to be abandoned temporarily while *Endeavour* sailed to the north-west to reach Tahiti in plenty of time to set up the gear needed for the observation of the Transit of Venus.

Endeavour was in excellent condition. Being so far away from a shipyard, Cook was very conscious that he had to care for every aspect of his ship, particularly the sails. As the wind strengthened, he would have the sails reefed to reduce the area of sail exposed to the wind. If a sail was completely reefed, it and the spar to

which it was laced, would be lowered to the deck to reduce wear and tear. For some time now, the seamen had been on 'watch and watch about', four hours on watch and four hours off. This ensured that there were always twenty seamen on deck, but it also meant that no one got more than four hours off watch, so sack time was at a premium. As soon as weather permitted, Cook resorted to a three-watch routine. This put fewer seamen on deck at any one time, but meant that, except for emergencies, each man had four hours on and eight hours off watch, a much less exhausting routine.

The marines spent much time on small-arms drill, so that they would be fully prepared, should their skills be required. Cook also had six of the carriage guns brought up from below and mounted them, three on either side, in the waist of the ship.

In the trade winds *Endeavour* plodded on at a stately five knots or so – 120 miles a day – the heavier sails were sent down, repaired and stowed below and a lighter suit of sails was put in their place. Rigging and anchor cables were examined, repaired or replaced when found wanting. The boats and their gear were checked and serviced. With Cook nothing was left to chance.

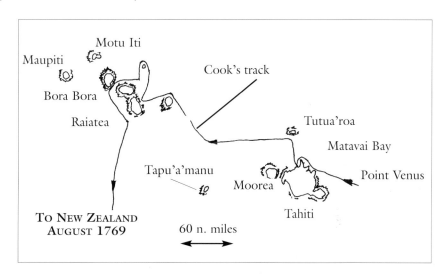

By now Cook had reached the reported latitude of Tahiti and was sailing westwards, until he met up with the island. This is a method known as 'running your westing down', or 'running your easting down', if you are going the other way. This was a normal navigational practice at the time, because finding the latitude could be done easily and accurately, whereas calculating the longitude was a chancy matter, as explained before.

It was not long before they began to see birds that were associated with land and they knew that they were getting close to a landfall. What they did eventually see,

Opposite: Saturday night at Sea, *an etching by George Cruickshank. Life on board was tough, dangerous and, at times, tedious. The men entertained themselves with songs, story-telling and a great deal of drinking.*

was one of the many atolls in the Tuamotu Group. The Tuamotus are also known as the Dangerous Archipelago, and for good reason. Atolls are always small and low-lying, often with off-lying reefs. Even in daylight they can be difficult to spot; on a dark night, they are impossible to see until you are too close to take avoiding action. When we were sailing from one of the Marquesas to Tahiti, we found that we needed to be within two miles of an atoll, before we could spot it, and that was in good visibility. True, that was from the deck of a yacht with a height of eye of no more than six feet, on a vessel the size of *Endeavour*, and if the lookout was way up a mast, then, of course, the atoll would have come into view that much sooner, but even so, it is all too easy to come upon one without warning.

Tahiti is a mountainous island and on 11th April, eighty days out from Tierra del Fuego, the tips of the mountains pushed their way out of the cloud-covered land. All landfalls are satisfying events for the navigator, but Tahiti must have had a special charm for Cook and his ship's company.

The high land disturbed the behaviour of the trade winds that fell away until *Endeavour* was becalmed. It took two more days to reach Matavai Bay, on the north coast. The anchorage was as good as Wallis had claimed. Fresh water was available from a river that flowed into the bay and the locals were friendly, almost too friendly, as their canoes crowded around the slow-moving vessel.

Endeavour approached Matavai Bay gently, urged on by a wind that was just strong enough to keep the sails asleep and the water chuckling along the side of the ship. The Tahitians were clearly welcoming, the brilliant coral sandy beach was backed by hundreds of coconut palms – it must have been a truly magical moment. Somehow, the huge anchor was dropped amongst the crowded escort of canoes, without sinking any. On board *Endeavour*, all would be activity, with some sails being furled and others backed, to send the vessel backwards and away from her anchor, in order to lay the cable out on the seabed. And then, at exactly the right moment, the cable would be snubbed to prevent it from running out further and the ship would continue backwards on her momentum, causing the anchor to dig in, until all was secure.

Wallis had been showered with stones when he left Matavai Bay, a year or so earlier, and Cook took no chances: the swivel guns were ready for action, as were the six four-pounders. The marines were drawn up, in full regalia, and armed, on the upper deck. Flags were displayed, wherever possible, to give an air of importance to the ship. Some prominent-looking Tahitians came aboard and it was obvious that they were willing to trade. They had learned the value of iron, principally in the form of large nails and spikes, when Wallis was in Tahiti. The Tahitian women were more than willing to offer sexual favours in return for an iron nail or two. Such a brisk trade was done during the month Wallis had spent in Tahiti, and his crew had extracted so many nails, spikes and bolts from the ships fabric, that its structural integrity was threatened! Cook did not want to risk this.

He was well aware of the Tahitians' craving for iron and had come fully prepared, with barrels of spikes, nails and bolts as articles of trade. In Cook's case, however, trade meant the purchase of breadfruit, coconuts, fish and pigs and not the favours of dusky maidens, no matter how glamorous! The trade goods were kept under a strict guard. Banks was appointed to the post of official trader, a task he carried out with fairness and some flair. A set of rules governing contact and trading with the natives had been drawn up by Cook, in which he insisted that the Tahitians were to be treated with friendship, fairness and humanity. In no time at all, however, it became apparent that the Tahitians, if friendly and welcoming, were also wily and highly competent thieves. Any member of the crew working ashore, who allowed his tools or arms to be stolen, would have the value of the lost items deducted from his pay and be awarded additional punishment, if deemed justified.

Similarly, anyone found to be trading with ship's stores, or with items made of iron, would be subject to the same penalty. This last rule was obviously made with Wallis's experience in mind. It must have been difficult to dissuade a sex-starved band of sailors not to use the Navy's property in order to buy themselves their little share of paradise. Captain William Bligh, who was to sail with Cook on his final voyage, writing of Tahiti after the Mutiny on the Bounty, said: 'The alurements of disipation [*sic*] are more than equal to anything that can be conceived.'

With the permission of the local chiefs, Cook had a defence post built, within which the observatory was established. The spot chosen was a sandy spit, bound on one side by the sea, and on the other by a wide river. Defensive trenches, six feet deep and ten feet wide, were dug on the two landward sides and the excavated soil was used to form a wall, four feet and six inches high. On the side flanked by the sea, a wall of earth some four feet high was built, with a palisade at the top. The remaining side was formed by the river bank, topped by a double row of barrels filled with sand. Two carriage guns and six swivel guns, that had not been mounted on *Endeavour*'s upper deck, were brought ashore and set up inside the stronghold. A tent was erected within the fort to accommodate the scientific instruments, one of which was a clock, its pendulum set at exactly the same length, as it was at Greenwich. Other tents were set up to provide shelter for those who would be staying ashore. Finally, the ship was warped into a position from which its guns could be brought to bear and supplement the ones within the defensive position.

Cook, never one to ride roughshod over the locals, or to take liberties with their possessions, would not allow a single tree to be cut down, before permission had been obtained and, if need be, compensation paid to its owner. Time and again, Cook's fairness and humanity shone, finding expression in his behaviour, be it with his ship's company, or the peoples of the territories he visited.

Some minor thefts occurred in the observatory and in the defence post and could be ignored, but, when the precious quadrant was taken, this could not be allowed to pass, as it was essential for the scientific observations about to take place and was irreplaceable. Banks and Green, along with a midshipman, set out in pursuit. After they had gone four miles or so, Banks sent the midshipman back to tell Cook what was happening. Banks and Green had to press on for another three miles before they came to a large crowd of mildly hostile Tahitians. Banks, who had two small hand guns, produced one and, at once, the crowd became less threatening. The quadrant was eventually returned, with some minor damage, which Mr Spöring, one of the scientists, was able to repair.

Despite the constant irritation of the petty thefts, Cook would tolerate no infringement of his rules regarding trade with and behaviour towards the Tahitians. As a result, a number of seamen were awarded twenty-four-lashes for breaking these rules, in one way or another. It was clear that cultural differences were bound to cause misunderstandings: some minor, some amusing and others more serious. The archery match between Gore and a Tahitian man illustrates the point quite nicely. It took both parties some little time to realise that their understanding of who the winner should be, differed considerably. Gore, of course, followed the western principle that the winner is the one who was most accurate. The Tahitian idea of an archery contest was to see who could shoot his arrow the greatest distance.

Tahitian men and women never ate together and found it repugnant when the officers of *Endeavour* invited women to eat with them. The men of the island actively encouraged their women to have sex with crew members, if it resulted in payment in iron, an attitude totally contrary to the beliefs of the Englishmen. With obvious cultural differences like these, it was only to be expected that there were others too, on both sides, that gave offence, without the offenders having the least idea of what had gone wrong.

Whilst waiting for the day of the Transit, Cook took every opportunity to go ashore, partly to maintain a friendly contact with the Tahitians, but also to observe their customs and way of life and the nature of the land in which they lived. Cook learned that Bougainville and his ship had visited Tahiti, not long after Wallis, and circumstances suggest that the French visitors left the population infected with syphilis. Cook did not know this at the time and he had all his men examined by the Surgeon who, with one exception, declared the crew free from infection. That man was not permitted ashore. Yet, it was not long before members of the crew were going down with syphilis, which indicates that the infection was a fact of life on Tahiti, before the arrival of *Endeavour*.

The day for the observation was approaching and Cook had set up two observatories: one was on the eastern coast of Tahiti on a headland now called Point Venus, and the other on the island of Moorea, about twenty-five miles away. The second observation post was set up as a precaution, in case the conditions in the vicinity of the main observatory made it impossible to take the necessary sights. This had entailed training some of the officers, petty officers and a midshipman to act as a back-up team, that could also take the necessary observations.

June the 3rd dawned with perfect conditions and the observations went off well at both stations. Unfortunately the results were less than conclusive. The object of the exercise was to record the exact time the edge of Venus came into contact with the edge of the sun, and, again when it left after crossing the face of the sun. The problem, that was to make the results gathered from around the world almost valueless, had not been foreseen. Both the sun and the planet Venus are surrounded by a penumbra, a hazy glow that Cook calls 'a dusky shade', which made it all but impossible to observe the edge of either body, with the required degree of precision.

Endeavour stayed for three months in Matavai Bay, and the period was used to give her a thorough overhaul. Stores were taken ashore, checked over and re-stowed aboard. Tropical sun and rain can play havoc with sail canvas and rigging, these were examined, repaired or replaced, where necessary. The anchors that had been down since they arrived, were brought on board. One by one, all were found to be eaten by worms, so badly in some cases, that they fell to pieces. The shipwright and his assistants had to find some suitable timber ashore and rebuild the damaged anchors.

Two marines were so enamoured of a couple of Tahitian women that, when the time came to leave, they decided to desert. They were soon rounded-up and taken back aboard, where they were punished, and they sailed with *Endeavour* when she left.

When the Tahitians realised that *Endeavour* was preparing to leave, a number of them wanted to sail with Cook, but he refused to take them. He argued that it would be unreasonable to submit them to the rigours of life aboard ship, especially as *Endeavour* was due to head south, into near polar conditions, nor could he accept the thought of them ending their days in the slums of London, or maybe as living exhibits in a freak show. Banks argued that a Tahitian pilot and interpreter would be a useful addition to the crew. In the end, he won the argument and a Tahitian, named Tupia, left with them. Tupia was a navigator and a politician who had been born and raised in Raiatea, a group of islands to the west of Tahiti, which Cook intended to visit, before turning south.

Bank's wish to take a Polynesian on board was not quite as altruistic as his argument would lead one to believe. His diary contains an entry, which suggests that he was well aware that the curiosity value of a man from the South Seas would be an asset to *him*, back in England. Tupia, however, proved his worth as an interpreter, in many of the places Cook visited, also giving him information on other islands. Sadly, Tupia only lived long enough to reach Batavia (the present-day Djakarta), where he contracted a fever and died.

In three short months, Cook and his crew had secured a place in the affections of the people of Tahiti. Cook's set of rules for dealing with the islanders was based on fairness and friendship, and his insistence that the members of the ship's company should not violate them, was of considerable importance in building a happy relationship. Banks ensured that the trading was conducted in a reasonable and civilised manner and that too must have helped build an atmosphere of mutual respect between the Tahitians and the British. On the few occasions when Cook felt it necessary to chastise some of the Tahitians, he did it with restraint. Indeed, on 13th July 1769, as *Endeavour* made ready to lift anchor, a large contingent of chiefs and other notables were ferried ashore, through a fleet of canoes manned by local people, all bewailing the fact that the ship was about to leave and beseeching Cook to stay. These images would stay in Cook's mind for ever.

Captain Cook's signature

7. Tahiti to New Zealand

July 1769 – April 1770

The Tahitians had told Cook of the presence of islands to the west of Tahiti and he resolved to visit them. As ever, Cook wanted to fill in a blank space on the chart, but he also needed to stock up with fresh food. Cook did not realise that the stock of fresh food in Tahiti was being badly depleted by the purchases made to feed the ship's company. Had they stayed much longer, there may have been problems and one can imagine that the population of Tahiti would be glad to see them go, on that score alone, despite the tears of the farewell party.

Tupia, the Polynesian who sailed from Tahiti with Cook, gave him the names of about eighty islands and sketched a map to give some indication of their positions. These ranged from the Marquesas, 950 nautical miles to the northeast, and Fiji, rather more than 1,500 nautical miles to the west. Tupia also gave Cook the sort of information about these places that could only be given by someone with first-hand experience of them. Following Tupia's directions, Cook found Huahine, the day after leaving Tahiti.

The chief of Huahine, an old man named Ori, developed an instant rapport with Cook and, as a sign of friendship, they ceremonially exchanged names. During his two-day stay Cook was able to buy some pigs and some vegetables to bolster their stock of fresh food for the next leg of the voyage.

Cook's curiosity about Bora Bora had been aroused when Tupia told him that the Chiefs on Tahiti and other nearby islands had long been in the habit to banish their criminals to Bora Bora, until the island became so overcrowded that it could no longer support the inhabitants. As a result, they turned to piracy. It seems that the Bora Borans raided Huahine on a regular basis, killing and stealing as they wished, and the good people of Huahine hoped that Cook might perhaps help them deal with the raiders. For Cook, Bora Bora was a disappointment in as much as contrary winds and extensive reefs made it impossible for him to get ashore. *Endeavour* sailed the area for a month, calling at eight islands whose people were clearly of the same stock as the Tahitians. He claimed all the islands for England and named them the Society Islands, in honour of the Royal Society.

Their next port of call was Raiatea. Tradition has it that Raiatea was the focus of the settlement of the nearby islands and was also the place from which a party left to migrate to New Zealand. Understandably, Cook wanted to visit it. This had been Tupia's home island, until some upheaval forced him to seek refuge in Tahiti.

Once Raiatea had been found, Tupia piloted *Endeavour* through the reef and into the protected lagoon. When Cook queried the depth of water, Tupia sent a local down to check, which dispensed with the need for a leadsman in the chains, as they approached the anchorage. The lagoon was idyllic, as indeed many of them still are in the South Pacific, even today. Tupia anchored *Endeavour* alongside what had been his ancestral lands, from which he had been forced to flee.

Cook stayed for a month at Raiatea to allow the summer season in the Antarctic to set in. It was going to be cold enough where they were headed, without going there towards the end of winter. All too soon, the time came to exchange the comfort of the tropics for the colder climes to the south.

Once outside the bay, Cook altered course to the south to resume the search for *Terra Australis Incognita*. Tupia told him that there were a few islands to the south, but that he had never heard of a large landmass in that direction. Nevertheless, Cook's orders were explicit, he was to prove or disprove the existence of the Great Southern Continent and the only way to do that was to sail those waters.

Four days after leaving the Society Group, Rurutu was sighted, but there appeared to be no suitable anchorage and the inhabitants seemed so hostile that Cook sailed on. He maintained his course for another sixteen days, until he reached 40° south and found himself in ever rougher and colder weather. He had been sailing westwards for some time, at the latitude stipulated by the Admiralty and all the indications were that there was no land for many miles. Albatrosses had been keeping them company and these giant birds are normally only seen far from land. For quite a while, the voyagers had been experiencing a large and hollow swell from the south and from Cook's experience that was a strong indication that the sea had travelled many miles without encountering land. *Endeavour* continued on this westerly course, in the same rugged conditions for a month, without encountering land, or even an indication of it.

By now the sails and the rigging were taking so much punishment from the weather that Cook turned to the north for a few days, seeking better conditions for ship and crew. As soon as the weather improved, he turned to the west, with perhaps some north in it, until he was on the latitude of New Zealand.

At last, on 5th of October, changes were observed. The colour of the water appeared paler, even though there was no bottom at 180 fathoms, the albatrosses were no longer with them, but Port Egmont hens and a number of other birds suggesting that land was near, appeared in their place. Changes like this do not appear without a reason and, sure enough, two days later, the boy on lookout at the masthead sighted land. *Endeavour* had reached the east coast of North Island, New Zealand.

There were a number of boys among the ship's company and the one who sighted the land, Nick Young, was inevitably known as 'Young Nick' throughout the ship. Cook, with his usual generosity, named the headland 'Young Nick's Head.' One can imagine the lad's delight. Young Nick's Head forms the southern point of Poverty Bay, so called by Cook because he was unable to get the fresh food he needed. At the northern point of this bay lies the present-day Gisborne, a town nestled against a backdrop of mountains, with the River Waipoua flowing into the bay. Banks felt sure that they had found the elusive Continent they had been looking for. Cook was unconvinced, but reserved his judgement until more evidence was available.

Over one hundred years earlier, the Dutch navigator, Abel Janszoon Tasman, had sailed in the vicinity of New Zealand and anchored in a sheltered bay, a little north of Poverty Bay. A party of seven men went ashore but, as soon as they were close enough, the Maoris attacked them. Four of the party were clubbed to death and the remaining three made their escape by swimming back to the ship. In view of this hostile reception, Tasman named the spot, Murderers' Bay, and moved on.

Cook was keen to get ashore and he headed for land in the yawl, together with a party of marines, closely followed by another armed party in the pinnace. Once the yawl was beached, Cook and some of his party stepped ashore, whilst the pinnace stayed back, close to the mouth of the river. Some Maoris approached stealthily among the trees, near the beach, and were clearly intending to isolate the yawl. The marines in the yawl were unaware of what was happening and the men in the pinnace shouted a warning to them, but it was not heard. Two shots were fired from the pinnace over the heads of the Maoris, but they ignored them. However, when one of the Maoris made to throw a spear at the crew of the yawl, he was shot by one of the marines in the pinnace. The Maoris were disconcerted to see their comrade dead, with only the sound of the gunshot to account for it, and they retired in double quick-time. At the sound of shooting, Cook and his party turned and raced back to the yawl and both boats beat a hasty retreat to the *Endeavour*.

When the Maoris first caught sight of the ship, they thought that it was a giant bird and that the human-like figures, dressed in brightly coloured uniforms, were gods. It requires very little imagination to equate this encounter with a spacecraft landing on earth today and one of the onlookers being killed by a 'space gun', whatever that might be. It must have been an unnerving experience for the Maoris, to say the least, and it is a measure of their courage that the next morning they were still on the river bank, ready to defend themselves and their village.

About fifty Maoris were waiting in full view of the ship, all of them armed with long spears and clubs, made of highly polished stone. Cook, Banks, Solander and Tupia landed on the beach, on the opposite side of the group of Maoris. The latter immediately became hostile, brandishing their weapons and breaking into a violent war dance. Cook fired a musket shot over their heads and moved his group back to await the arrival of his escort of marines. When they were assembled ashore, the party moved to a point opposite the Maoris. From there Tupia called to them and to everyone's surprise he was able to communicate with them. There seemed to be enough similarities between the Tahitian language and that of the Maoris to make themselves understood. The Maoris wanted to know where the strangers came from and complained about the killing of the previous day. Tupia eventually managed to convince them that the Englishmen meant no harm and that the man had been shot in self-defence. Eventually, they agreed that food could be bartered. Not surprisingly, it was an uneasy truce: the Maoris refused to put down their arms and would not cross the river to meet the party from *Endeavour*.

The stalemate was finally broken when Cook left his weapon with one of his party and walked down to the river's edge by himself. One of the Maoris entered the water and swam to a rock in the middle of the river, where he waited for some time before

Opposite: *The head of a Maori warrior, as observed and drawn by Sydney Parkinson. The European visitors were fascinated by the tatoos, the clothes and ornaments worn by the islanders. These artefacts were drawn with just as much care as the flora and fauna of the territories visited by the* Endeavour. *Parkinson, however, followed the neo-classical conventions of the period and the faces of his subjects display features that are more European than Maori. Also see the portrait of Poedooa on page 209.*

completing the crossing. Cook offered him some small gifts. A second man swam across, trying to conceal the fact that he was armed, hesitantly, the remainder of the group followed. Small gifts were distributed but this did not satisfy the Maoris for long, Nothing was safe from them, but their real interest lay in getting their hands on some of the Englishmen's weapons. Despite a warning, that anyone attempting to steal a weapon would be killed, they continued their efforts to appropriate some of the arms of the ship's party and, eventually, one of them succeeded in gaining possession of a knife. Cook ordered the marines to open fire. The thief was mortally wounded and three others were injured.

With two dead and three wounded, Cook's hopes of making friends with the locals were beginning to look less and less likely, but there was worse to come.

Endeavour was approached later in the day by two canoes. Still hoping to win the locals' trust, Cook decided to change his tactics. The men in the canoes appeared to be unarmed and he decided to capture one or two, take them on board, treat them well and, having gained their trust, send them back. When Tupia spoke to the group, they paddled away from *Endeavour*. Hoping to gain their submission by frightening them, Cook ordered a gun to be fired over their heads. Instead of backing down, the Maoris turned and treated the ship to a barrage of large stones. Clearly these men were of a different calibre from the Tahitians. This hail of stones was dangerous enough to prompt Cook to order the marines to return fire. Three or four Maoris were killed, and at least one other wounded. The three who had escaped injury dived into the water and were picked up by seamen from *Endeavour*.

Not surprisingly, the three youngsters were terrified, but they slowly calmed down, as they realised that no harm was intended them. They eventually explored the ship, had a meal with the crew and stayed the night, seeming to enjoy their experience. What surprised everyone was their lack of concern for those of their companions who had been shot.

Considering the fatalities of the previous days, the officers agonised for much of the night over the disastrous results of their attempts to make friends with the Maoris and over the best way to handle the situation in the coming days.

It was Cook's intention to put the boys ashore, where he had landed the day before, but they would have nothing of it. They claimed that the people who lived there were enemies and, given the chance, would kill and eat them. Whilst Cook was inclined to doubt the stories of cannibalism, he fell in with their wishes and had them put ashore on the far side of the bay.

There seemed little reason for Cook to stay in Poverty Bay and, as soon as the boat that had taken the three lads ashore was back and hoisted inboard, the anchor was raised and *Endeavour* got under way. Very soon after, a number of canoes came out from the shore and their occupants indicated that they wished to board *Endeavour*. These were friends of the three youngsters who had stayed on board overnight and who had been convinced by them of the Englishmen's good intentions. They were there to satisfy their curiosity and to trade. Once the nature of the visit had been established, Cook had the anchor lowered and the sails stowed.

The visitors were particularly keen to obtain lengths of 'Tahitian cloth' of which *Endeavour* had a stock. This paper-like cloth is called 'tapa' and is not woven, but made of the inner fibres of the bark of the paper mulberry tree. The bark is soaked in water and when it is ready for the next stage it is laid on a fallen tree trunk where it is beaten with a wooden club. The beating separates the inner fibres from the outer bark and it is carried on until the fibres are matted together to form a sheet of stiff fabric. The final step is to decorate it boldly with vegetable and mineral dyes. The manufacture and use of tapa was not confined to Tahiti, it was found throughout Polynesia. Although the present day Polynesians use modern textiles, it is still possible to hear the tap, tap, tap of the clubs at work in some parts of Polynesia. As in Cook's day, a part at least of the production, is sold to tourists! There is a melancholy footnote to this. Cook brought back a length of tapa home to his wife. As she waited for him to return from his last voyage, she set about making and embroidering a waistcoat for him, made of the Tahitian cloth, a garment which Cook would of course never have the chance to wear.

A few of the friendly groups of Maoris stayed the night on board *Endeavour* and the next day they encouraged more of their kinsmen to join them. Being many probably made them feel secure and, as Tupia was able to converse with them, it soon became clear that the fear of strangers and of possible cannibalism was at the root of the hostility displayed by the first people they had met.

Cook and his colleagues were struck by the similarities between the indigenous population of New Zealand and the Tahitians. Pre-eminent was Tupia's ability to talk with them with little difficulty, even though he came from an island, some 2,000 nautical miles away. The similarities did not stop there: culture, artefacts and so on had so much in common, that the only sensible explanation was that the Maoris had, at one point, migrated from the east. To a seaman, this would be the logical route, in as much as the trade winds and the ocean currents move from east to west in these latitudes. No sailing man in his right mind fights the elements, he uses them, especially if his craft is primitive. Yet there are still 'experts' who insist that the migrations in the Pacific went the other way, ie against wind and current. I can only presume that they are not seamen and wish that they would go and try it! When my wife and I were sailing from the Galapagos to the Marquesas, it took us three weeks sailing *before* the wind. We were aware of the simultaneous efforts of a family who was attempting to circumnavigate the world *against* the winds. They sailed from the Marquesas, headed for the Galapagos. They had to give up after 142 days and return to the Marquesas, unable to fight the winds.

It is now known that in the 200 years spanning the end of the eighth century and the beginning of the eleventh, migratory voyages were undertaken by the Polynesians from perhaps as far east as the Marquesas. These voyages of some 2,800 nautical miles brought Polynesians to New Zealand and a number of other places. It is believed that there was a further migration from Polynesia in the fourteenth century. The settlers established themselves in the North Island of New Zealand, where they flourished, and eventually spread to the whole of New Zealand. By the time of Cook's visit they had increased in numbers to over 100,000. The migrants had to make changes to their lifestyle to allow them to live in a harsher climate than they had been accustomed to, but so many cultural similarities remained that were strong enough and varied enough to banish doubts as to their origins.

When Cook left Poverty Bay, he turned to the south to examine the coastline, and hoping to carry out a circumnavigation of New Zealand. He was also anxious to find fresh food, this entailed finding a suitable place to anchor and to put men ashore, but the rugged coastline offered little or no shelter. From time to time canoes would bring parties of Maoris offshore. They seemed driven by a mixture of curiosity and determination to prevent the alien intruders from landing.

With the deaths of a number of Maoris still fresh in their minds, the English were reluctant to appear hostile, unless forced to defend themselves. On one occasion a group of Maoris came close enough to attempt to trade. Tupia's servant boy, Tayeto, leaned over the ship's rail, was seized and taken off in the canoe. Shots were fired and, in the confusion, the boy was able to escape and swim back to *Endeavour*, very frightened, but otherwise unhurt. Two or perhaps three of the Maoris were killed, and more would have died, but for the danger of shooting the boy. This incident caused Cook to name a nearby headland, Cape Kidnappers.

Having resumed his southerly course it was not long before Cook encountered head winds and, naturally, the further south he went, the colder it became. It made sense to turn about and head north with the winds, rather than fight them. As the summer season was about to begin, it also had the advantage of leaving the exploration of the more southerly regions until the weather had warmed up a little. The run south had taken Cook unknowingly to within fifty miles of the south-eastern point of the North Island, before he reversed his course. The point at which he headed north, he named Cape Turnagain.

As *Endeavour* headed northwestwards, the natives they met were friendlier and more willing to trade with them. One group of Maoris actually guided the British ship to a small bay, where water was available and the holding (for their anchor) good. The crescent-shaped bay proved to be quite beautiful and the people friendly. The attitude of these Maoris made it possible for parties to go ashore and collect wood, water and some fresh food. Cook was especially pleased to find that two anti-scorbutics: scurvy grass and wild celery, were growing in profusion nearby.

Scientific parties were also able to go ashore and, with the aid of Tupia who was popular with the locals, they were able to establish a useful rapport with them, even visiting them in their homes and observing their way of life. Cook called this anchorage Tolaga Bay, a name it has kept to this day, its meaning in the Maori language being simply 'a landing place'.

On 29th October *Endeavour* left Tolaga Bay and sailed northeastwards until they reached the easternmost point of New Zealand, which Cook named East Cape. The land they could see from the ship looked fertile and productive and the area seemed to be more heavily populated than the other places they had seen. En route to East Cape canoes of Maoris had taken to the water but they were unable to reach *Endeavour*. On one occasion, however, a number of canoes approached the ship with what seemed to be hostile intent and Cook ordered a round of grapeshot to be fired ahead of the canoes, which was enough to scare them off.

While most of the canoes were of modest size, they saw one which consisted of two canoes, each about seventy feet long, lashed together, side by side, with athwartships beams to form a very powerful and seaworthy catamaran. This vessel carried about 200 men, outnumbering *Endeavour*'s crew by more than two to one. The catamaran sailed alongside *Endeavour* for some time, while the Maoris talked with Tupia, but, when some manoeuvre on board the English ship was misinterpreted as hostile, the Maoris let fly with a well-aimed broadside of large stones. Cook had his marines respond with a volley of musket fire, whereupon the catamaran was taken out of range.

To many explorers, the finding of new territory was an end in itself. To Cook it was only a beginning: once it was found, it had to be catalogued in as much detail as possible. Cook would record the latitude and longitude of islands and reefs that lay offshore, capes and other points of interest on the coast and prominent features ashore, such as mountain tops and the like. Captain Cook's logs abound in information that would be of great value to other mariners for years and years to come. The availability of fresh water, wild celery and other anti-scorbutics at the anchorages he used, the degree of shelter to be found and from which winds, the quality of the holding and the depth of the water at points around the coast, were all noted painstakingly.

From the middle ages until the middle of the 1900s, seamen calculated the period of the high tides, from the time the full or new moon crossed the meridian of a particular place. For a number of reasons these high tides – High Water – arrive some time after the passage of the moon and this time lag is called 'the establishment of the port'. Cook was always careful to note 'the Establishment of the port' of the places he visited. Whenever possible, he also noted the heights to which the tide would rise at high and low water. It was only around the 1950s that Admiralty charts ceased to have a note of the hours and minutes of 'the establishment of a nearby standard port', printed in the margin.

Another matter of considerable importance to mariners is called 'variation'. This is the difference between true north and magnetic north. If there was no variation, the magnetic compass needle would point directly to the geographical North Pole. Unfortunately the earth's magnetic field is not uniform at every point and the magnetic pole is not at the geographical North Pole, but somewhere in north America. What is more, it wanders about a little. Clearly, if the compass needle is treated as if it were pointing to the geographical North Pole, the navigator could miss his landfall completely. Fortunately, an observation of the sun made at sunrise or sunset and subjected to a simple calculation, will give the variation for that spot and allow the navigator to correct his compass course. Cook never failed to record the local variation in his log.

As Cook sailed along the coast of North Island, he could see that the Maoris had built pallisaded defensive positions, close to most of the villages, which suggested that the locals were not at peace with one another. *Endeavour* made her way along the coast, stopping to collect fresh food, record salient details and, at all times attempting to establish friendly relations with the local populace. The ship's log shows that on 8th and 9th November, the Maoris brought out a considerable quantity of a kind of mackerel, which was more than enough to feed all ninety men of the ship's company.

Thursday, 9th November, was the day when the Transit of Mercury was to be observed, this, like the Transit of Venus on Tahiti, would be of value to astronomers back in England. Whilst Mr Green observed the movement of Mercury, Cook took the altitude of the sun to enable him to calculate the 'apparent time' as part of the observation (apparent time is the same as sundial time).

Whenever possible, *Endeavour* sailed by day and, if it was a moonlit night with good weather, they would also sail through the night, but generally they anchored at night. If the place merited it, the ship would stay on her anchor for a few days. Most of the time, the Maoris were friendly, even, on some occasions, welcoming, but on 30th November, Cook, Solander and Banks went ashore with armed parties in both the pinnace and the yawl. Very soon after they had set foot on the beach, they were surrounded by two to three hundred armed men. They did not, it seems, appear to be an organised group of warriors, although they were armed, but simply a large hostile crowd. Soon, some of them attempted to seize the two boats. The men guarding the boats discouraged them firmly and they turned their attention to Cook's party. Cook, Banks and two of the marines fired their muskets at those at the front of the group. The muskets had been loaded with small shot so as to avoid causing serious injury, and it was not enough to make the Maoris retire completely: they came back to face another volley of small shot.

Some officers and men aboard *Endeavour* observed the disturbance. They quickly rigged a spring (rope) off the main anchor cable back to the ship's side, paying out some more anchor cable. This caused *Endeavour* to turn broadside to the Maoris on shore and allowed the gunners to fire five four-pound balls over the heads of the hostile crowd, causing them all to retreat in a great hurry.

By early December *Endeavour* was anchored in the beautiful Bay of Islands, but, as soon as they had replenished their fresh water and taken on board a large quantity of wild celery, the anchor was weighed and they headed north-west yet again. The next day they were plagued by head winds which held for the best part of a week, forcing them to tack back and fore, making only limited progress. They then encountered gales that drove them back, causing them to lose the ground they had fought so hard to gain. On 16th December the weather relented, and they were off again to the north-west, but it didn't last long. The wind came in strong from the westerly quarter and they had a long and hard slog trying to round North Cape. On 21st December the winds drove them seventy-two miles north of North Cape, and the battle went on until the beginning of January, when they finally rounded the Cape. The wind still blew from the west and the vessel was now in danger of stranding itself on a lee shore, which prevented the close examination of the coastline.

After a few days, the wind turned to somewhere between north-east and north-west, enabling them to come close to the shore safely. *Endeavour* then continued on her way to the south-east, buoyed by a variety of winds from light to gale force. The moment the wind turned onshore, the course would be altered to take the barque offshore and away from the danger presented by the lee shore.

On 4th January there is an entry in Cook's Log that will pull at the heartstrings of all blue-water yachties. On one occasion, he sailed close to a beach that had large waves breaking upon it, when, without warning, the wind swung round and became an onshore wind and the fight to claw themselves off began. In his log, once clear of the danger, he swore that he would never approach such a dangerous coastline again, without a favourable wind. Amen to that!

Having found myself in a similar situation, albeit in a vessel that could go to windward more readily than *Endeavour*, I know just how tense he must have felt until his ship was safely offshore. I know what it is like to be responsible for one or two other lives. To know that a wrong decision could cost ninety or more lives, with no hope of help from any quarter, is a disturbing thought!

The exploration and survey of the New Zealand coast continued in this vein for six months. During that time Cook circumnavigated both the North and South Islands, surveying and recording details that would allow the production of charts of the area on his return to England.

They were no longer in the tropics and often there was no sun to illuminate the underwater dangers for them. Bays in which to anchor had to be found and checked out, before they could attempt the hazardous business of sounding their way in. Whenever possible, the botanists would be landed to carry out their tasks and Cook himself would go ashore to make notes on anything that might be of value. He often noted the suitability of the standing timber for ship building. On one occasion, for example, he noted trees that grew to eighty feet high, before the first branches appeared – perfect for making replacement masts. On another visit, he found Maoris cutting timber for the construction of a large canoe. He measured a plank that was prepared and ready for use, it was some seventy feet long by two inches thick and twelve inches wide. Making such plank would have been a considerable task for English shipwrights, equipped with the best hand tools available at the time, but for the Maoris working with their stone tools, it was a remarkable achievement.

The scientific party had no background in or real understanding of navigation and ship-handling problems. On occasion, they were apt at asking Cook to facilitate their work, by undertaking some manoeuvre that no prudent seaman would contemplate. Clearly, there were times when mild discord arose on this score. The example most frequently quoted was the request made by Banks and company, for Cook to take his ship into a 'harbour' that was in the nature of a narrow mini fjord, running east and west. Westerly winds were plentiful, so there would be no problem getting in. Getting out, on the other hand, would require an easterly wind. Cook had noted that in the area easterlies blew on fewer than one day a month and he refused to take his ship into a foolhardy situation. He was also undoubtedly aware that the bottom of fjords are often rocky and the water very deep, giving little hope of anchoring safely, if at all. To a seaman, the place reeked of danger. Cook categorically refused to go in, which did not stop the scientists from insisting.

One can imagine the aristocratic Banks being badly put out by this refusal to carry out his request and enable him to carry out his work, but even *he* had to accept that Cook was the captain of *Endeavour* and that his word was law. That he resented it bitterly is evident from comments in his journal and, even as much as thirty years later, he was still angry at what he regarded as lack of co-operation on Cook's part.

On 27th March there is a note in Cook's log, indicating that the circumnavigation of New Zealand is now completed and clearly his thoughts turn to leaving. He needed wood and water before the long journey home and *Endeavour* was taken into Queen Charlotte's Sound and anchored close to the shore. The shipwright and his party were sent ashore to cut and transport wood, while others collected some thirty tuns of water. By the evening of the 31st, the ship was ready for sea again and at daylight the next day, the anchor was brought aboard and *Endeavour* was under way, heading westwards.

So much of what Cook did was remarkable, but the circumnavigation and survey of the coast of New Zealand is one of his greater achievements. To complete the task, Cook had to sail two and a half thousand miles, most of it not far from the shore, the most dangerous place a ship's master can place his vessel. The information he brought back was of such a quality that the charts produced from it were probably the best work Cook had done so far. The charts in use in the early 1940s were still based on Cook's initial survey. He had made only two mistakes of any importance: on one hand, he thought that Banks' Peninsula was an island, an understandable mistake in as much as the land connecting the headland to the South Island is very low-lying. On the other hand, he listed Stewart Island as a peninsula, at a time when he had to contend with very tough weather conditions and poor visibility.

In 1943 the American armed forces were based in New Zealand and the Sea Bees (US Construction Battalions) decided that these charts should be brought up to date. They carried out a survey of the New Zealand coastline and its nearby waters. The survey brought to light several apparent discrepancies in Cook's work. Some years later, when satellite surveying was introduced, it revealed that Cook had been correct and the US Sea Bees, wrong!

Another of Cook's remarkable achievements was that after a two-year voyage, there was not a single case of scurvy on board. Louis de Bougainville, who had left Tahiti some ten months before Cook, was barely able to sail his ship into Batavia, as his crew was decimated by scurvy. By a strange coincidence, Captain Jean François de Surville, aboard the *Jean Baptiste*, was attempting to circumnavigate the northern end of New Zealand, in the opposite direction from Cook, at exactly the same time. At some point, their ships were only some thirty miles apart. By then, sixty of Surville's crew had already died of scurvy, and he had only been at sea for eight months. Surville is often regarded as the very opposite of Cook. The latter's impeccable seamanship, his concern for the safety of his ship and health of his crew contrast sharply with Surville's disregard for prudence and the poor care he took of his men who were exhausted and near starvation. He nearly lost his vessel in New Zealand's severe winds and offshore weather. In one gale he lost anchors, cables and a dinghy. If Cook's encounters with the Maoris were not always of the friendliest, Surville's were a disaster. Cook, always concerned to increase his store of knowledge and with charts in mind, liked to sail close to the coast, Surville stayed away from land, whenever possible. He made a short reconnaissance of New Zealand, then sailed east – this time with the prevailing winds – only to drown while trying to get ashore in Peru.

Octopus

8. The Queensland Coast

April 1770 to July 1770

Cook took his departure from the north-western corner of New Zealand's South Island on 1st April 1770. He named it Cape Farewell. Cook had been away from England for two years and had done all that had been asked of him. He could sail east and home via the Horn, or he could go westabout and home by the Cape of Good Hope. The easterly route would entail getting well down to the south to pick up the east-going winds. Cook knew that this would expose him and his ship to the extremely strong westerlies and heavy seas that are to be found in those latitudes, especially during the winter season which was now getting into its stride.

Two years of hard sailing had taken its toll on *Endeavour's* gear and she really was in no fit condition to face a long spell of seriously adverse weather. The alternative was to sail westward from New Zealand towards Australia, then turn north to follow the east coast of Australia to Cape York. From there he could make the East Indies, cross the Indian Ocean, round the Cape of Good Hope and head north in the Atlantic, until he reached home. This route would give *Endeavour* and her ship's company much kinder conditions. The only difficult part in this passage would be the rounding of the Cape of Good Hope, better known to seamen as the Cape of Storms. In the circumstances, it was the seamanlike decision.

Banks was not enamoured of the idea; not for the first time it was the scientist versus the seaman. He wanted to continue the search for the Great Southern Continent, in spite of the onset of winter. He was willing to concede that Cook and others had covered so much of the area to the south, that any unknown continent had to be a great deal smaller than the library-bound geographers had suggested, but he was not yet convinced it did not exist.

Cook was very conscious of the welfare of his ship and his people. He knew that the spars had been weakened by storms. Sails and rigging were aged and weakened by long exposure to the tropical sun and were daily creating problems. Rigging was parting and sails splitting, sometimes blowing out of their bolt ropes in shreds and the sailmaker was fast running out of canvas and sewing twine. They were also running out of anti-scorbutic agents and Cook feared that he would be unable to renew his supplies in the latitudes he would need to take his vessel to, if he decided to make use of the westerly winds. As much as he might have been prepared to go on looking for the Great Southern Continent, it was no longer a practical proposition.

Banks and Cook debated the possibility of a second voyage, with the completion of the search for the Southern Continent, as one of its main goals. By the time *Endeavour* reached home, Cook had organised his thoughts on the subject and was ready to attempt to persuade the Admiralty and the Royal Society that a second Pacific voyage of discovery would be a worthwhile project. When Cook reported to the Admiralty on his return to London, he found that he was knocking on an open door. For the present, however, Cook was primarily concerned with Australia.

It is probable that the first European explorer to see Cape York, the north-eastern tip of Australia, and some of the north coast, was Luis de Torres in 1606. Because the Spaniards kept the details of his discovery of the strait that now bears his name, secret for so long, the rest of Europe did not know if Australia and New Guinea were part of the same landmass, or were separated by the sea. Cook wanted to settle this question. Dalrymple had managed to obtain a copy of a document outlining Torres's achievements and he included details of the Torres Strait in his pamphlet, entitled *Discoveries made in the South Pacific, Previous to 1764* and published in 1767. Perhaps out of spite towards Cook, he presented a copy of it to Joseph Banks to take on the Transit voyage and it would fuel the running argument between the latter and Cook over whether the strait really existed.

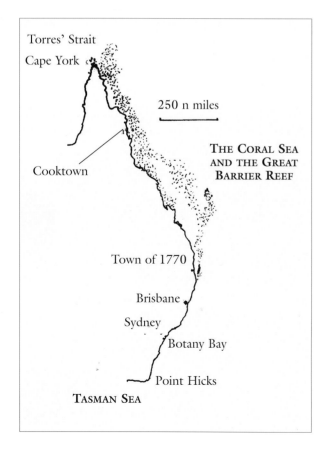

THE CORAL SEA AND THE GREAT BARRIER REEF

Ten years after Torres's visit, a Dutchman named Dirk Hartog, found the west coast of Australia and over the next several years he and other Dutchmen explored the area until it was fairly well charted, along with a section of both the north and south coasts. With a firm foothold in the lush East Indies, just to the north, the Dutch thought of Australia as barren and unproductive and lost interest in it.

The eastern coast of Australia was mostly virgin territory as far as Europeans were concerned. It is not surprising that this should have been so. Between it and the Coral Sea, lies the Great Barrier Reef, which stretches from just north of Brisbane to New Guinea, some 1,200 miles or more. At its narrowest point the Reef extends forty miles into the Coral Sea, but some 150 miles at its widest section. The whole of this area is a mass of closely packed reefs, cays and islands which, in places, reach to the coast of Queensland. The Tasman Sea, between western New Zealand and the east coast of Australia, can be home to some pretty foul weather. Winds funnel up from the Antarctic, that are both cold and strong. A fact which is sometimes overlooked, is that cold air is heavier than warm and, hence, a gust of cold air striking a sail at thirty knots will do more damage than a similar one of warm air, simply because it is heavier. Air does not come much colder than that from the Antarctic and *Endeavour* suffered. It was not all gales between Queen Charlotte's Sound and Australia, but she had more than her fair share and the wear and tear was considerable.

Two weeks after leaving New Zealand, Cook's reckoning told him that he must be getting close to his landfall. So, on the night of 17th April, instead of continuing westwards, he had *Endeavour* sail short boards (tacks), first to the north and then to the south, to the north again and to the south, and so on until daylight. By this method *Endeavour* remained within easy distance of her landfall, without running the risk of striking the land in the dark. This time-honoured tactic is still used by blue-water yachties, who are approaching a landfall without navigation lights, once darkness has fallen.

The next day, their feeling that they were approaching land, was reinforced, although it had not yet been sighted. Land birds were spotted, the distance run from Tasmania was about right and the longitude was close to that given by Tasman for 'Van Diemen's Land' as he had called it.

Late afternoon on the next day, Cook reduced sail and at 0100 he brought to and had a leadsman sound the depths. Using 130 fathoms of line they found no bottom. *Endeavour* continued her cautious approach under reduced sail until daylight, when the reefs in the topsails were shaken out and, an hour later, land was sighted by Lieutenant Hicks. Cook named the spot Point Hicks. Captain Cook was no fool, and I can't help wondering if his apparent generosity in naming places after the first man to sight it, was perhaps his way of ensuring that a good lookout was maintained.

Today, there is a Point Hicks on the southeast extremity of Australia, close to the latitude and longitude given by Cook. Unfortunately, there is some ambiguity in the

Overleaf: The finished painting of Banksia serrata *by John Frederick Miller and based on a sketch, accompanied by notes and colour references, by Sydney Parkinson. The actual specimen, shown on the left, was collected at Botany Bay in early May 1770. It was an important addition to the science of botany, being a new genus. The plant was called Banksia after Joseph Banks who found it.*

THE QUEENSLAND COAST

figures given, which casts doubt upon the exact point on the coast Cook actually named Point Hicks. As there is a rather specially shaped headland, not far to the east of Point Hicks, which Cook named Ram Head, 'because it resembled Ram Head [now Rame Head] close to the entrance to Plymouth Sound,' it would seem that Point Hicks cannot be too badly misplaced.

From this landfall, *Endeavour* was put onto a northerly course, to make her way up the coast we now know as New South Wales, and further north, to Queensland. The large, hollow seas that rolled in from the south-east and broke heavily on the beautiful beaches, forced Cook to keep his distance from the shore and caused him to miss the exploration of harbours like Jervis Bay and even of Sydney.

Some of these beaches are today the haunt of surfboard enthusiasts but, they were, and indeed still are, far too hazardous to approach in a vessel, relying solely on the wind for its motive power. Cook was finally able to approach a bay, a few miles south from where Sydney stands today, and find an anchorage in six fathoms, at the southern end of what was first called Stingray Bay. The name was later changed to that of Botanists' Bay, which has become Botany Bay. It afforded meagre protection, but at the southern end, *Endeavour* must at least have been sheltered from the everlasting rollers. There was no bar at the harbour entrance to cause them to ground on the way in and, when they entered the bay, the wind was blowing out to sea, which augured well for their departure.

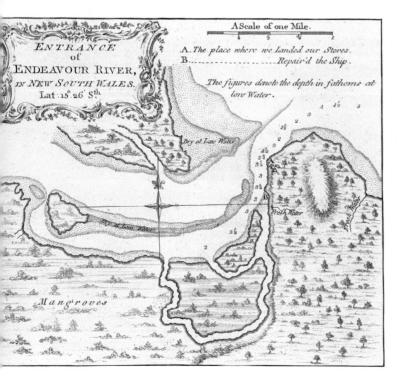

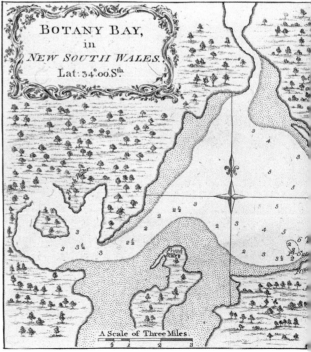

The botanists, along with the wood- and water-parties, were put ashore where they started on their work. Attempts were made to establish contact with the Aborigines, but they did not respond.

The scientists were delighted. The rich soil yielded an abundance of unknown plants and trees. Within a few days, Banks collected so many specimens, that he and his assistants were hard put to preserve and catalogue them, before they spoiled. The fauna was equally rich and fascinating.

The flag was raised and the appropriate words were spoken to ensure that the whole of the continent now belonged to His Majesty King George III, but still no contact had been made with the inhabitants of this vast country. On May the 6th, having replenished his water supply and his stock of wood, Cook had the anchor raised and stowed and *Endeavour* was on her way north again. The weather was kind to them, the wind coming from a direction that allowed Cook to sail close to the coast. As they proceeded north the land became hillier, with fine sandy beaches and occasional outcrops of rocky headlands.

Ashore, the lush growth turned into tropical rainforest, but this stretch of coastline lacked shelter for shipping. *Endeavour* continued to sail north for three weeks, making a single stop to replenish the water supply. Quite suddenly, they found that they were in shallow water and surrounded by reefs and small islands. They had reached the southern end of the Great Barrier Reef. Little did they know that it covered such a vast area.

The ship's boats were put into the water and their crews sailed ahead to find a safe passage for *Endeavour*. Cook was keen to continue the survey he had embarked upon from his first contact with the Australian coast and went on sailing close inshore, to enable observations to be taken.

Before darkness fell on 22nd May 1770, *Endeavour* was anchored two miles NNW of what is now known as Round Hill Head. The next morning a party, that included Cook, was put ashore in Round Hill Creek. Cook's arrival is celebrated by two

Opposite: Charts of Endeavour River and Botany Bay, *1770. The results of Captain Cook's surveying work during his first voyage included these two charts. Botany Bay was so named because of the great abundance of flora and fauna found there in May 1770. In June of that year,* Endeavour *spent several weeks at Endeavour River, where the damage it had sustained when it struck a reef could be repaired.*

monuments, as well as by the presence of a tiny township, named Seventeen-seventy, which developed on the spot where Captain Cook first set foot, in what is now Queensland. When we anchored there ourselves in 1986, it was almost as peaceful as it must have been, when Cook landed there.

Endeavour continued on her way north, generally sailing by day and anchoring at night. There were occasions when it was possible to sail by night. I suspect these were nights when the wind was gentle, but strong enough to give good steerage way to the ship, and when there was a cloudless sky and a good moon. Throughout the passage, the time, bearings, cross bearings, sun-sights, tidal observations, water depths were taken and recorded, and the magnetic variation was calculated for use in chart-making, when Cook eventually returned to England.

The Queensland coast is an area of outstanding beauty. Cook was sailing between the coast, with its backdrop of tropical rainforest on the port hand, and the nearby reefs and islands on the starboard hand. In some places, the reefs come very close to the coast, for Cook, sailing without charts or prior knowledge, the beauty of the area would obviously be tempered by concern over the potential dangers posed by under-water hazards.

Whilst *Endeavour* was making her way north, leadsmen would be in the chains sounding, and, if the water began to shoal, the ship's boats would be sent ahead to check that there was enough water for the vessel to float safely. When the boats were not employed in checking the depth of water, they would have been used to explore and survey the islands and reefs, lying off to the east of the shore, adding to Cook's store of information. Even so, they would only have penetrated the landward edge of this immense reef.

Shark

Opposite: Abelmoschus moschatus. *Finished painting by Sydney Parkinson, made during Cook's first voyage, 1768-1771.*

9. Endeavour Strikes a Reef

11th June 1770

On 11th June 1770, *Endeavour* struck a reef and was damaged below the waterline. There should be no surprise that Cook ran his ship up onto a reef, the amazing fact is that he had not done so sooner, surrounded as he was by this immense collection of islands, cays and coral reefs. When we sailed in these waters ourselves, we used over seventy Admiralty charts, *he* had nothing.

The minute coral polyp that built this vast area of reefs and islands needs sunlight to survive, so there is a limit to the depth that coral can be built. With the aid of the rise and fall of the seabed, and of the sea itself, over many thousands of years, the coral has built some remarkable slab-sided reefs and islands. As an experiment, scientists drilled a hole through the coral, down to the bedrock of an island in the Pacific. The hole is one-mile deep! The point I wish to make is that one can be in very deep waters one moment and in dangerous shallows the next, and if the light is wrong, there is no warning. This is what deceived Captain Cook. When we sailed into Flying Fish Cove, Christmas Island, we had our echo sounder set to read from zero to one hundred meters. As we approached the cove, the water was too deep for the echo sounder to give a coherent signal and then, quite suddenly, it read 2.5 meters… so we anchored.

For a seaman deep water offers safety. There was no way Cook could have known just how far the Barrier Reef extends to the east. Accordingly, at about 1800 and in preparation for the coming night, Cook shortened sail and altered course to ENE which would take him away from the shore. A clear moonlit night was in prospect and the weather was good. In those conditions it was perfectly reasonable for him to continue sailing. By 2100 the depth of water had increased from fourteen to twenty-one fathoms but, quite suddenly, it was reported as shoaling, first to twelve and then to eight fathoms. The depth began to increase again, eventually reaching twenty-one fathoms. The leadsman continued to report depths of this sort until 2300, when

Opposite: *This engraving, probably based on a drawing by Sydney Parkinson, shows* Endeavour *beached for repair after striking the Great Barrier Reef.*

it dropped to seventeen and, before the leadsman could make another cast, *Endeavour* had struck. Immediately, there was a flurry of disciplined activity. For some time now, the crew had been at their stations, ready to drop the anchor in case the water shoaled dangerously. But the grounding happened so quickly, that there was no time to take measures of that kind. The sails were lowered and stowed, boats were launched to sound the depths around the ship and to find where deep water lay. In places, the depth was as little as three or four fathoms, and in others, as much as eight to twelve. Where *Endeavour* had struck, there was barely three to four feet of water. Her momentum had carried her some distance onto the flat top of the coral. She required an absolute minimum of two-and-a-half fathoms (fifteen feet of depth) in which to float. By comparison with other places, the rise and fall of the tide is small in the area where she grounded. To make matters worse, *Endeavour* went aground on a falling tide and, even a small fall

makes it impossible to move a vessel from that position. *Endeavour* was there to stay, at least until the next high water, and maybe longer.

Meanwhile, the crew was busy lightening ship by casting over the side any heavy items that were not needed immediately, or would not spoil in the water. Many things were buoyed so that they could be recovered later. Sails and spars were brought to the deck and then put over the side; six of the guns were buoyed and put into the water; stone ballast followed, as did any and every thing that could possibly be spared.

Two kedge anchors were put out, one to starboard and another right aft, and were hauled taut. But it was all to no avail, more water was needed before *Endeavour* could hope to come free of the reef. All night, the ship's company laboured to lighten her and by 1100, the next forenoon, the time of high water, they had put forty to fifty tons of gear into the water. Despite these efforts, when high water arrived, it was found that a further foot of water was needed to let the vessel float off. Until now, they had taken little or no water, but, as the tide fell away, *Endeavour* began to leak badly. It was necessary to keep two pumps in continuous action to get rid of the water.

In the course of the afternoon of the 12th, the two bower anchors were taken out and laid, one aft and the other on the starboard quarter. That short sentence covers a considerable amount of very hard labour and a great deal of intelligent seamanship. The bower anchors are a ship's main anchors and they are *big*. To lay an anchor in these circumstances, it would need to be lowered, so that it was all but submerged. Two large pulling boats would then be ranged, one on either side of the anchor, then lashed together with spars, laid across them to form a raft, with the anchor hanging between the boats from the athwartships spars. The 'raft' was then rowed to the place where the anchor was to be laid. At this point, the lashings holding the anchor were released, allowing it to drop to the seabed. Windlasses were then used to haul the cables taut, so that if the water rose high enough and *Endeavour* floated, she would not drift further onto the reef, but could be hauled off by heaving on the anchor cables.

By 1700 the tide was clearly rising and, as it rose, *Endeavour* took in more water. A second and a third pump were brought into action, the fourth and last pump refused to function. At 2100 the ship lurched to an upright position but was still held fast by the lack of water over the coral. The leak increased, gaining on the pumps at an alarming rate. Cook decided to attempt to haul his ship off, by heaving in on the anchor cables. All hands, that could be spared from the pumps, were turned to the windlass and capstan. At 2220, *Endeavour* came free and floated into deep water.

The boats set about retrieving the smaller anchors and the hands, who had been working on the capstan, were sent to help the men operating the pumps.

Early in the forenoon watch the pumps had gained significantly on the leak and some members of the crew were set to sewing small lengths of oakum and wool into a lower studding sail which would then be used to fother the damaged area. Fothering involves covering the damaged section of a ship's side or bottom with a sail to reduce the influx of water. If there is time, as there was in this case, short tails of oakum, caulking cotton, wool etc can be added to thicken the layer of canvas, rather like a rag rug, increasing the chances of plugging the hole in the vessel.

The recipe worked in this case and, once the sail was lowered over the damaged area and secured in place with a series of lashings that went right round the ship, the leak was reduced to the point where it could be controlled by a single pump. The next step was to make their way to the coastline, some twenty miles distant, find a sheltered spot where *Endeavour* could be careened and repaired. The crew started to get the spars and sails etc back aboard, hoisting them into position to allow some sail to be raised. By noon on 13th of April, they were about nine miles from the shore. Boats were out ahead of *Endeavour*, both sounding and looking for what Cook calls 'a harbour'. On 14th they were still blessed with mild weather and were quite close inshore, when, in the afternoon, the boats found a suitable inlet. Around the same time the wind piped up. The haven seemed to be just what was wanted, but the entrance channel was narrow, with shoals on either side and, with a rising wind, Cook was forced to anchor off and wait for some improvement in the weather.

Friday the 15th brought a southeast gale. While Cook waited for a reduction in the wind strength, he had the top gallant yards, the fore top gallant mast, the jib boom, the spritsail yard and a number of sails in the fore part of the vessel lowered to the deck, as a first step towards lightening *Endeavour*, as soon as they were able to enter the haven.

At 0600 on Saturday 16th, the wind appeared to be moderating, so they hove short on their anchor cable to be ready to take advantage of any reduction in the strength in the wind. Unfortunately, it was only a temporary lull and they had to pay out the cable again.

Sunday 17th brought a lull and, this time, they were able to take advantage of it and get under way. They stuck on the sand bar that ran across the mouth of the river but, as the tide was rising, they soon came off and headed in for shelter. Again, they struck the sandy bottom and came to a halt. It was not a serious problem, however, as it happened on the weather shore and the wind would eventually blow them off, once the tide rose sufficiently. They, in fact, did not manage to free themselves until the next day.

When *Endeavour* floated off, she was warped alongside a steep beach to allow for work to begin. The fore yard and topmast booms were brought to the deck and lashed together with the other spars that had come down earlier, to form a raft, on which the shipwrights could work.

By Friday 22nd June, the main part of the damage to the hull was accessible and, although serious, it was repairable. The shipwrights, armourers and others worked away to put things to right. Meanwhile other members of the crew were busy collecting water and fresh food. This work continued into the first week of July, by which time *Endeavour* was ready to have her spars and sails brought aboard and put back in place.

A month later, the barque was ready for sea but the wind came from the south-east, which made it impossible to sail out of their haven. Whilst *Endeavour* lay at her anchor, work parties made use of the unavoidable delay, seeking fresh food, fishing or hunting turtles.

An anchor had been laid on the seaward side of the bar so that, when the time came, they could warp themselves out of the mouth of the river. They were wind-bound until Saturday, 4th of August, when at 0500, the wind dropped away to a calm. By midday, they were over the bar and out of the river, now called the Endeavour River. Today, a small township, called Cooktown, has developed close to the spot where *Endeavour* was repaired. The reef that did the damage has been named Endeavour Reef. The names chosen by Cook himself for salient points along that coast: Cape Tribulation, Weary Bay and Hope Island, say a lot about the frustration he must have felt.

Once out of the river, they anchored in fifteen fathoms, over a sandy bottom. At this point on the Queensland coast, the Great Barrier Reef comes in very close to the mainland and continues to do so, right up to Cape York. Cook wanted to survey the shoals and reefs at low water from the masthead, before deciding on his best route northwards.

They were fortunate, in as much as they had good weather with gentle breezes throughout this operation. Poor weather could have made things very difficult, if not impossible. Had the ship been lost, the death toll would have been high, as there were not enough small boats to accommodate the whole crew and they were too far offshore to swim to safety. Cook and his men found themselves on the coast of Queensland during the southern winter, had it been the summer months, they could well have have been struck by a tropical revolving storm. These storms routinely have winds of 120 to 150 knots, and, in the circumstances, few or none would have survived. Cook's seamanship and leadership played their part, but his crew was equally able and willing. In his journal, Cook speaks very highly of their behaviour during this emergency. Thanks to Hollywood, the popular image of a mutinous ignorant rabble, manning the sailing vessels of those days is, generally speaking, quite unfair. The tasks that had to be carried out by the crew of a sailing ship required craftsmanship, a deep sense of responsibility, and a dogged determination to get the job in hand done, qualities that all too often remain unrecognised.

A storm

10. Towards Batavia and Home

August 1770 to July 1771

The view from the masthead was not encouraging. Cook and a number of his officers had spent a lot of their time up there, searching for a safe passage, but they appeared to be surrounded by breakers and, where there are breakers, there are shoals. Their position was a difficult one. Cook discussed this with his officers and the Master. Some of them and the Master favoured turning back and heading south. Cook's instinct, I feel sure, was to find a way northwards into the unknown, rather than covering ground he had already explored.

Doubtless, he also had a better understanding of the difficulties involved in turning south than those who advocated this route. To go south would mean beating a thousand miles against strong headwinds, followed by something like an eight-thousand-mile haul across the Pacific. If Cook retraced his steps, the logical route would be to pick up the westerlies, south of New Zealand, and sail straight for Cape Horn. The winds in this area are very strong and they can kick up formidable seas. Then there was Cape Horn, with all that it can mean to a weary sailing vessel. The crew were up to it. Among their contemporaries they were probably the healthiest ship's company to have travelled so far, but finding enough fresh food to keep them well, on the long leg from Australia to the Horn, via the southern route, would be difficult. *Endeavour* had sailed a long way already and gear and materials do not last for ever. To have exposed her to the rugged conditions to be expected in the westerlies, would have been unwise.

Before they could decide on the route home, they were faced by a more immediate problem: *Endeavour* started to drag her anchors. This put the vessel in a dangerous position. The culprit was a very strong southeast wind and the nearest shoals were only about two miles away, downwind. A great deal of cable was let out on both anchors, it slowed the drag, but did not stop it. The windage aloft was reduced to relieve the anchors of some of the strain. The top gallant mast and several lesser spars, complete with their sails, were lowered to the deck and this halted *Endeavour*'s drift downwind, towards the reef.

At last, on Friday 10th of August, the wind dropped sufficiently for them to get under way. Cook had decided to take the northern route. The pinnace was out ahead sounding the channel and lookouts were posted at the masthead. The following day, things were so uncertain that they anchored and Cook himself explored the surrounding waters in the pinnace. He landed on an island, about fifteen miles to the east of the mainland, stayed overnight and, the next morning, climbed the highest hill so that he could view the surrounding waters to some distance. The pinnace had been out sounding the depths and had found a couple of narrow channels, but the strength of the wind made it impossible to examine them properly.

Because the island had a large population of lizards, many three or more feet long, Cook named it Lizard Island. He also commented on the fresh water that was available a little way inland. Both water and lizards were still there when we visited Lizard Island. Some 'modernisation' has taken place since Captain Cook's time: the stream had been directed into an underground cistern and a shower rigged above it, with a hand pump to raise the water. The rain forest shrubs growing around the shower area provided us with a measure of privacy. Fine, but the plumber has arranged things in such a way that the pump was just beyond the reach of the person under the shower!

The irony is that if Cook had continued to sail parallel with the coast, within a few hundred yards, he would have found a perfectly good channel to take him all the way to Cape York. He wasn't to know that, of course, and he sought safety by sailing away from the coast.

With Cook back aboard, *Endeavour*'s sails were raised and, with the pinnace sounding ahead, they made their way to the north. There is a channel, now named Cook's Passage, that runs through the reefs from Lizard Island towards New Guinea. This must have been the route Cook found and used to escape from the difficulties that had beset him for so long, but not without one final brush with the Great Barrier Reef.

In the small hours of Thursday, 16th of August, they had breakers ahead, stretching to their left and right, as far as they could see. The wind was such that they felt confident they could sail parallel with the reef, until they found a way round it, but very soon the wind changed direction and started to push them directly towards the reef. They changed tack and sailed to the south, which took them obliquely away from the danger. *Endeavour* made about two miles on this tack, before the wind dropped altogether. The water was too deep to allow *Endeavour* to anchor and this left her to the mercy of the tides that run strongly in the area. By dawn they could hear the roar of the surf, they were that close! As the sun chased the night away, they could see the heavy seas breaking on the reef and throwing a cloud of spray, high into the air.

The boats were hoisted out in an attempt to tow *Endeavour* away from the danger. A cat's-paw of wind came in and helped the boats move the barque, but it soon fell away again, returned, then dropped to a calm once more. The wind continued to play with them for some hours, but all the time the ship was getting closer to disaster, in spite of the best efforts of the men in the pulling boats. At just about the last moment, a light breeze sprang up and stayed long enough to enable them to coax *Endeavour* into a lagoon within the reef, which was shallow enough to let them anchor, but sufficiently deep to allow them to float. It had been a close-run thing.

The circumstances were quite different than those of the previous incident, but, although they suffered no damage on this occasion, potentially, it was much more serious. It really was the sailor's ultimate nightmare: to find himself stuck aboard a ship, without a breath of wind with which to work it, while it is slowly being carried to a violent end and having no means of anchoring, rendering him totally helpless, with all his nautical skills of no use whatsoever.

Saturday brought gentle breezes from the right direction and they made their way south from the lagoon and headed northwest, bypassing the vast reef and sailing back towards the mainland. Despite the difficulties and dangers, Cook and Green, throughout this latest problem and the stranding on Endeavour Reef, as this ill-fated spot became known, never failed to take their sights each day and to record their position with the greatest of accuracy.

Cook's next task was to establish whether or not New Guinea was joined to Australia and this was one reason why he had been keen to get back to the mainland. By 22nd of August he was in the vicinity of Cape York, the northeast corner of Australia. The Torres Straits lay before him, and the proof that New Guinea and Australia were not joined as a single landmass, was his.

Endeavour continued west, eventually leaving Booby Island, that outpost of Australia, astern, and making her way through the Arafura and Timor Seas, mostly with favourable winds, but with the occasional hiccup, when the water shoaled dangerously.

On 29th August the masthead lookout spotted land to the north. Cook waited for daylight and approached the land cautiously. This was the Dutch colony at Timor. *Endeavour* coasted along until Savu came into view. Cook and others went ashore to attempt to buy provisions from the people who could be seen in a village close to the

Opposite: *Booby Island in the Timor Sea*

water's edge. The locals were keen to sell, but the Dutch officials were far from helpful. The Dutch were very protective of their trade monopoly with the East Indies and they went so far as to refuse any form of help to passing mariners. Cook already knew about Dutch officialdom, but the lure of fresh food was ever present.

By Monday, 1st of October, they realised that they had overshot the western end of Java and turned back, with the intention of calling at Batavia. This Dutch possession had been founded in 1619, on the ruins of Jakarta. It was a fortified city which had rapidly become the main port in the area, the seat of Dutch power in the East and the headquarters of the Dutch East India Company. The port was like a little piece of the Netherlands, transplanted into a steamy tropical landscape. Cook and his men had endured a tedious passage on their way to Batavia but, when they eventually arrived, it was to find that their problems were yet to come, thanks to the frustrating process of dealing with Dutch officials. Not only did Cook need fresh food, but *Endeavour* was in need of some serious repairs, before she tackled the rest of the

journey home. It was believed that most of *Endeavour*'s hull was in good condition, but she was leaking at the stem and around the area where she had sustained damage. It is probable that both areas were leaking for the same reason: a considerable strain must have been transmitted to her stem when she hit the reef.

The stem is the foremost vertical timber in the vessel and the planking from either side is fastened to it. The stem sweeps down to meet the keel, to which it is fastened, and it is usually curved through something approaching a right angle. It is too much to expect to be able to cut such a massive piece from one single piece of timber. Normally, two or more pieces are scarphed together. The scarph in *Endeavour*'s stem was opening, creating a leak and weakening the vessel. The problem had to be dealt with promptly as these things never get better of their own accord.

Cook wanted permission to enter the area where the Dutch dockyard was set up, careen his vessel and carry out the repairs. A permit was needed for everything and the committee that gave them out sat once a week or so. The least alteration to the approved work had to be presented before the next sitting, and so on, *ad nauseam*. Whilst they were waiting for permission to proceed with the repairs, the crew were occupied with preparing the vessel to be hove down. Permission was eventually secured, but on condition that the Dutch shipyard carried out the repairs.

Work started on Friday, 9th of October, and when *Endeavour*'s underwater damage was revealed, it proved to be far worse than expected. A large part of the deadwood was gone. The copper sheathing, which had kept the teredo worm at bay, was missing over large areas. The dreaded worm had attacked some eighteen feet of planking and, in some places, only one-eighth of an inch of the original thickness of the hull timber was left. The keel was also badly damaged in a number of places.

The steamy climate and the unhygienic conditions of Batavia were notorious for their detrimental effect on the lives of those who went there. When Cook arrived, only two members of his ship's company were mildly sick: Tupia and Mr Hicks. Quite soon Mr Monkhouse, the Surgeon, died of a fever and this marked the start of a period of ill-health that had not been experienced before aboard *Endeavour*. By mid-November, Cook had difficulty in mustering more than a dozen men, fit for work.

The Dutch authorities ashore, seemed to care little for matters of public hygiene. Sewage flowed in open drains to the sea. The hot sticky climate encouraged flies, mosquitoes were abundant and, in such circumstances, widespread and serious sickness was almost inevitable. Some symptoms displayed by those who fell ill suggest malaria and what was generally known at the time as the 'bloody flux' – probably dysentery. It was perhaps a tribute to the general fitness of *Endeavour*'s crew that not more than seven men were lost to whatever malady attacked them in Batavia. But the sickness was to stay with them and others died on the passage home. On the day they left, the sick numbered forty, with the rest in very poor shape. There was one exception: the sailmaker. Aged between seventy- and eighty-years of-age and more or less drunk every day of his service, *he* remained untouched by anything except alcohol, at least for the moment. If there is a moral there, I am not quite sure what it is!

It took three painfully long months to get the repairs done and the ship made ready for sea. *Endeavour* left Batavia, early on the morning of 26th December 1770, but did not get very far. The wind was flukey and then died away altogether, making it necessary for them to anchor. For the next ten days they were plagued with winds that did not serve, or died on them and, consequently, they spent a lot of time on their anchor. Unfortunately, they were at the southern edge of the doldrums and could expect nothing better until they had worked their way a little further south.

Eleven frustrating days later, *Endeavour* had covered a mere 300 miles and was anchored off the south-east corner of Princes Island. As the locals were willing to sell fresh food of all kinds, Cook decided to stay a few days, using the opportunity to replenish his water and wood stores as well.

They left to a gentle breeze that soon died and they were forced to anchor yet again. The sick were not helped by the sultry, windless conditions. There were several more deaths, the alcoholic sailmaker amongst them, even he was not immune. Cook had the ship cleaned with vinegar, in the hope of banishing disease.

January wore its weary way on, with disappointing daily distances. For a fortnight *Endeavour* averaged runs of twenty-four miles a day. In their worst twenty-four-hour period, they made a mere four miles. Nevertheless, all the time *Endeavour* was inching her way southwards and, finally, at the end of the month she began to work her way out of the doldrums and pick up winds that enabled her to make some worthwhile daily runs.

Early in February, the winds strengthened and those men who were not too sick began to recover slowly. Sadly, there were others who were too far gone already and by the end of February a further twenty-seven men had died. Cook had worked so hard to maintain the health of his crew, his journal merely records the bare facts, but the loss of so many men must have hit him hard. Luckily, the weather was kind, which was just as well as, at times, so few crew were in a fit state to work.

They reached Cape Town on 16th of March, dogged all the way by sickness and death. An early priority was to get the sick ashore, in the hope of saving as many as possible. The Surgeon was sent to find suitable accommodation for them but, in spite of the improved conditions, there were to be further deaths.

Endeavour left Table Bay in mid-April 1771, the right time of the year to give her a kindly passage home. In one six-day period she made good 855 miles – an average daily run of 142 nautical miles – which was good going for her. It is quite exhilarating to sail along in this manner: the wind unchanging day and night and for days on end, the ship surging forward with each roll of the ocean, a marvellous sensation that has to be experienced to be understood. Undoubtedly, it must have had its effect on the sick men and helped speed their recovery.

It was not only the crew that was in poor health. *Endeavour* was weather-beaten and battered. Almost daily, something would give way or a sail would snap out of its bolt ropes, with a crack like a gunshot, but no matter, as they were nearly home. Good seamen have always prided themselves on arriving home in a spick-and-span ship, the Royal Navy is no exception. I am sure that *Endeavour* was scraped, holystoned, painted and polished to the nth degree before they reached the coast of Britain. In the western approaches they were met by a southwesterly gale that fairly chased them up the Channel, in a welter of spray and spume, and no doubt, Cook and his Master spent much of their time looking anxiously at the gear, wondering what would give way next. Yet, the old collier must have looked quite something as she passed the familiar headlands, with every pennant flying, the great ensign at her stern outshining them all.

With no more fuss than when she left three years previously, *Endeavour* anchored in the Downs, just off Dover, on 13th of July 1771. Cook left his ship and, in the company of Joseph Banks and Daniel Solander, went off to report to the Admiralty, where he received a very warm welcome. He handed over his logs, journals and the charts he had made of his work in the South Pacific. He also gave detailed reports on the anti-scorbutic measures he had established, as well as on the performance of the navigational instruments that had been entrusted to him.

James Cook was promoted to the rank of Commander, which was really not very much of a promotion, considering all he had achieved. In today's Navy, the rank of Commander is a substantial one, suggesting that the recipient is likely to rise further. This was not the case in Cook's day when it was often a sticking point in an officer's career. The rank above Commander was that of Post Captain and it is to this rank, that one would have expected Cook to be promoted, at this stage in his career.

To be without a ship in Cook's day meant loss of pay. To avoid this, he was appointed to command the *Scorpion*, which was moored in the Thames. He was presented to the King and generally fêted within the Service, where his contribution to matters nautical was greatly appreciated, if not quite as well rewarded as it might have been.

That the appreciation of Cook's efforts was genuine, is clearly shown by the fact that it was decided that another Pacific expedition under Cook's command, should be organised as soon as possible. Cook had been working on the aims of a second voyage, long before he got home, and he was well pleased to find himself pushing at an open door.

The press of the day showed surprisingly little interest in James Cook and his achievements, but Banks and his fellow scientists received enough publicity and adulation to more than make up for Cook's lack of mention. Much of what was written by the press regarding Bank's work was pure fiction. For example, he was credited with bringing specimens of the laurel back from the Philippines, a country he never visited. It was also claimed that he had discovered the Great Southern Continent – which does not exist – and that he had carried out the observations for the Transit of Venus, for which Cook and Green were responsible. Later, it was also reported in the press that Banks was to have two ships from government sources to enable him to mount another expedition to the South Seas. A few months later, it was announced that the celebrated Mr Banks was to have three ships at his disposal. None of this was ever denied or corrected!

At that time, the press also carried a story to the effect that, when *Endeavour* grounded on the reef, a hole was punched in her side. Fortunately, a huge lump of coral had allegedly been torn from the reef by the force of the collision and had plugged the hole in *Endeavour*'s hull, thus preventing her sinking. Later editions went on to describe in great and lurid detail how the shipwrights stood, chest-deep in the water, to drive oakum into the spaces between the coral and the ship's timbers to keep the water at bay. All fanciful nonsense, I fear. It seems that journalism does not change. Today no disaster at sea fails to occur in anything but 'mountainous seas and shark-infested waters!'

Alexander Dalrymple, in common with other persons of privilege, had been allowed to read copies of Cook's journal and to say that he was not pleased, would be to put it mildly. You may recall that when the Royal Society had proposed a voyage of exploration of the South Pacific, they had chosen Dalrymple to lead it. The latter was greatly offended when the Admiralty turned him down. Dalrymple had been one of the proponents of the *Terra-Australis* theory and Cook had gone some way towards demolishing the idea, which did nothing to endear him to Dalrymple and his cronies. They took a vindicative pleasure in belittling Cook's work, or in claiming that the discoveries had been made by someone else. Dalrymple claimed that the proof of this allegation was to be found in his book, published before Cook set out on his voyage.

It is true, that before Cook left England, a certain amount of information on the geography of the South Pacific was available to seamen. But it ranged from the useful to the fanciful, and a good deal of it owed its origins to wishful thinking, rather than to factual observation. This material was used by self-declared scientists, who had never seen the Pacific, and by others who should have known better, to bolster their case for a Great Southern Continent.

Before you think that such a thing could not happen in modern times, give a thought to Thor Heyerdahl's magnificent ocean voyages, made aboard replicas of ancient crafts, the *Kon Tiki* and the *Ra,* that sailed at ten knots and could be steered. His detractors insisted on describing his sea passages as 'drifts', rather than as voyages. The publication of the accounts of these expeditions brought a storm of protest and ridicule from those who considered themselves to be experts in the field of pre-Columbian archaeology and history. Heyerdahl's work was belittled and scorned by people who had built their scientific reputations, on at least partially wrong thinking, or misinformation. I am not aware of one expert who, at the time, had the courage to say: 'this man's approach is worth looking at.' Thirty years later, changes have taken place in the understanding of pre-Columbian history and of archaeology, that can be tracked back directly to Thor Heyerdahl's work. Although scant credit seems to have been given him, at least Heyerdahl lived to witness the scholars' change of heart. Cook was not so fortunate.

Whilst we are on this tack, it might be the place to mention coconuts. It had long been claimed that coconuts floated from Asia against the wind and current, please note, and came ashore at numerous points in the Pacific to take root and grow into fine trees. As an experiment, Thor Heyerdahl left a large number of coconuts soaking in sea water for three months, by which time they were no longer viable. Marine borers had attacked some and the rest had absorbed sea water through the 'eyes' and rotted.

When we were in Australian waters, we learned that there had been no coconuts in Australia, until the settlers brought them from elsewhere and planted them. Cook's

Opposite: *The author's yacht,* Didycoy, *sailing.*

journals confirm this – he would never have missed the chance to collect coconuts. It was not until he reached the region of Cape York, where coconuts could have been carried in native craft from New Guinea and planted in north-eastern Australia, that he records obtaining some to add to the crew's diet. And yet, we are still being told by those who should know better, that coconut palms were spread around the world by coconuts that had floated across the oceans.

11. Preparations for the Second Voyage

September 1771 to June 1772

Little less than two months following Cook's return to England, the Admiralty was actively looking for a ship for a second voyage to the Pacific. Despite the wear and tear to which she had been subjected, *Endeavour* was still seaworthy. She had been refitted and sent to the Falkland Islands with stores for the settlers.

Mindful of his two close brushes with disaster, Cook recommended that this time there should be two ships to sail in company, and the Admiralty readily agreed. It made good sense.

We know that Cook was involved in the selection of the two vessels that were to make up the second expedition. He had been well satisfied with the performance of the Whitby collier, it had suited its purpose and he aimed to use the same kind of vessel for the new voyage. After some searching, the Admiralty found two colliers that had been built to ply their trade in and about the North Sea and purchased them in mid-November 1771. Both were similar in design to *Endeavour*. When they were taken into naval service they were renamed *Drake* and *Raleigh*. *Drake*'s 450 tons made it larger than *Endeavour*, while *Raleigh*, at 336 tons, was slightly smaller.

It occurred to someone at the Admiralty that both Sir Francis Drake and Sir Walter Raleigh had been thorns in the side of the Spaniards. It was thought that it might perhaps be more diplomatic to change the names of the vessels to something less provocative, just in case Cook required Spanish goodwill, if not assistance. As a result, the two ships became *Resolution* and *Adventure*. Because of the publicity surrounding the return of *Endeavour*, there was no shortage of volunteers to crew the two vessels, many of them using whatever influence they could bring to bear to improve their chances. One lad in particular was to carve out his own place in history, fourteen-year-old George Vancouver joined *Resolution* as an able seaman. Later he was to command his own vessel and carry out exploration and survey work in the Pacific for the Admiralty. Vancouver Island bears his name. Cook was given a large measure of autonomy over the choice of his crew members. At least twenty of those chosen by him had served aboard *Endeavour*. Joseph Gilbert, who had worked

Opposite: *The Map of the Southern Hemisphere drawn by Joseph Gilbert.*

with him on the Newfoundland survey, was appointed Master of *Resolution*. Six of the thirteen seamen from *Endeavour* who were taken on were promoted to the rank of petty officer. Perhaps the most interesting appointment made by Cook was that of the marine, Samuel Gibson, who was one of those who had fallen in love and attempted to desert in Tahiti. A lesser man than Cook might have excluded the marine for his past misdeeds, Cook enlisted him and then promoted him to corporal.

The botanical and natural history side of the voyage was again to be in the hands of Joseph Banks. Cook and Banks have always seemed to me to be a strange combination of characters. Cook was a modest man who had worked his way up from low down in the social scale, whilst Banks was definitely upper crust and flamboyant with it and, yet, the two were good friends, especially aboard, where they shared the same dangers and tribulations. When ashore, the differences of background and education drew them apart inescapably. It has been said that Banks, in particular, had been on an adventure, while Cook was doing his job. This does not detract from Banks's scientific contribution, but refers to their individual motivation for the journey. This time Banks planned on taking an even larger party, at least fifteen strong – not counting their servants. The group would comprise botanists, physicians, artists, draughtsmen and two musicians, whose role would be to entertain the gentlemen.

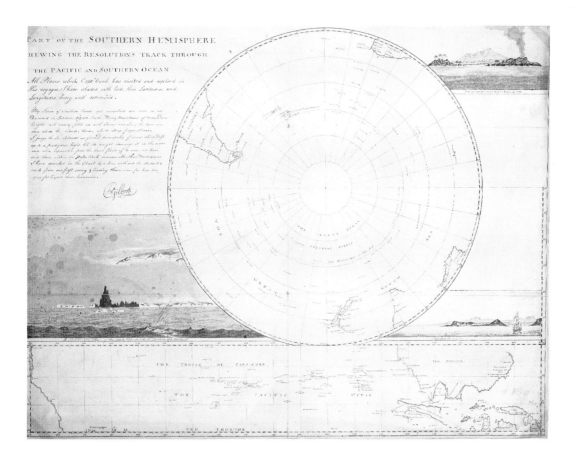

Banks had basked in the attention lavished on him since his return and his ego, never in short supply, had been inflated to an even higher level than normal. Public opinion needed a dashing hero and Joseph Banks was only too pleased to oblige. Cook worked on grimly, reporting on his past voyage and beginning to organise the second, while coping with personal grief. When he had finally made his way home to Mile End, it was to discover that his fourth child, Joseph, whom he had never seen, had lived for only a few months and that his daughter Elizabeth, his third child and a favourite of his, had also died. Besides these personal tragedies, he also had the grim task of writing to the families of the men who had perished during *Endeavour*'s voyage.

Banks, on the other hand, was the toast of the town. He, who had failed in his studies at Oxford was now being awarded an honorary doctorate and his showy accomplishments put Cook's great geographic and scientific discoveries almost in the shade. The great Linnaeus himself, suggested that New South Wales should be renamed *Banksia* and that a statue to the 'immortal Banks' should be erected. On the other hand, Cook hardly received a mention in the press. It was at the Admiralty, and among his sponsors, that his truly remarkable qualities were being appreciated. At that time, he certainly wasn't the one for grand dinners and receptions, or for courting press attention.

Banks made it known that on that second voyage he would expect to travel on something grander than two ex-colliers. A major warship, or the sort of vessel used in the East Indies trade, would be far more to his liking.

He was horrified when he went to see what he regarded as *his* ship. It afforded barely enough space for his own luggage, forgetting his companions'. When he realised that Cook and the Admiralty had made their choice and were not going to change their minds, he asked them to modify *Resolution* to increase the accommodation space for his group. Anticipating a refusal, he turned to his friend, Lord Sandwich, and gave him an ultimatum. *Resolution* would be altered as he wished, or Banks would abandon the voyage before it had even begun. Lord Sandwich was conscious that it would not look good if the Navy was seen to be preventing the famous Banks from carrying out his all-important work and he gave his consent. Although there were some dissenting voices, the Admiralty Board agreed to the alterations and work was put in hand. Accordingly, the deck of *Resolution* was raised by twelve inches from the quarter deck to the forecastle, a new cabin was built for Cook, as the party would occupy the captain's old cabin, as well as the entire Great Cabin.

Alterations of this kind were bound to add a great deal of extra weight high up in the ship. It would raise the centre of gravity of the vessel with likely serious consequences for its stability. Banks's power of persuasion must have been great for his demands not to be turned down on these grounds alone, but James Cook must have been dismayed at what was happening to his ship.

The provisioning of the ship was usually left to a junior officer to oversee, but Cook undertook this task himself. His success in almost banishing scurvy aboard *Endeavour* and in maintaining a good standard of health among his crew, until he called in at Batavia, disposed the Admiralty to acquiesce to his requests. Fifteen tons of sauerkraut and three tons of so-called 'portable soup' were supplied as part of the anti-scorbutic diet. Portable soup was made of a decoction of vegetables, mixed with liver, kidney and other offal and boiled to a pulp. Dehydrated and hardened into slabs, it had to be soaked in water, to which were added cooked dried peas, oatmeal being sometimes used in place of the peas. Large quantities of malt and a syrup made from oranges and lemons were also taken aboard. Cook was asked to test the efficacy of carrot marmalade in the prevention of scurvy. He also undertook to carry beef that had been preserved in various ways, in order to see how well it kept over a period of time.

Benjamin Franklin and Alexander Dalrymple urged the Royal Society to supply stocks of livestock and seeds to be taken to New Zealand. The animals were to be released and the seeds sown, in the hope that there would be ready supplies of food for future visitors. The plan fell through because the Royal Society once again proposed that Dalrymple should be given command of one of the ships and lead the expedition. The Admiralty's opinion of Dalrymple had not improved since he had striven to be given the leadership of Cook's first expedition in 1768. The answer was unchanged, there was no way that the Navy would allow Dalrymple to command one of their ships. Despite the Royal Society's withdrawal, Cook did carry a large quantity of seeds and livestock obtained from other sources.

Apart from Joseph Banks's work on all things natural, a great variety of scientific research and of testing of new instruments was to take place on this second voyage. For instance, James Cook was asked by the Commissioners of the Navy to test a new type of azimuth compass. He was also to try out two methods of producing fresh water from sea water: one was by distillation and the other by injection of carbon dioxide. Why these two particular experiments needed to be conducted at sea, is difficult to understand.

More importantly, Cook was also required to test four chronometers. If they remained accurate in the taxing marine environment, they would greatly simplify the determination of longitude at sea.

Astronomers of the time could calculate the longitude of a place ashore by a method that required them to find the 'lunar distance'. The lunar distance can be described as the angle between the moon and the sun, or the moon and a star, whose position was listed in the astronomical tables of the day. This was not an easy observation to make ashore, afloat it ranged from very difficult to impossible. This observation, plus a couple of other factors, had to be subjected to a rather complex piece of mathematics and, with a bit of luck, the answer was your longitude. This was fine in principle, but even ashore it was subject to error and it was usual to have two or three observers take simultaneous sights, so that a certain degree of personal error could be averaged out.

Afloat, it was a different matter. The difficulty of taking accurate observation from the heaving deck of a rolling and pitching vessel, rendered the operation difficult indeed. A suitable star – ie one already listed in an almanac such as the volumes published by Dalrymple and his colleagues – and the moon, had to be visible together. Alternatively, the moon had to appear when the sun was available. Clearly these juxtapositions did not occur every day. The technique involved was also lengthy. Even in ideal conditions, it could take four hours or more to complete the observations and work the calculations. The astronomer could afford to wait for the right situation. But when a navigator needs to know his position, he generally wants to know it *now*, not in three or four days' time. He also needs to be able to rely on its accuracy and, as we have seen, the difficulty of making a lunar distance observation at sea could easily introduce considerable errors into the calculation, thereby leading to an erroneous position. Errors of forty miles or more were not unusual. For those ship masters who lacked the necessary observational and mathematical skills, the technique was quite beyond them.

In 1714 an Act of Parliament was passed that offered £20,000 as a prize for the inventor of the first workable method of finding longitude at sea, to an accuracy of half-a-degree. A second prize of £15,000 would be paid for a method accurate within two thirds of a degree and a third prize of £10,000 for accuracy within one degree. The errors permitted could be as great as sixty nautical miles. These tolerances were so large as to make the resulting answers practically worthless, but if they were regarded as an 'improvement', it says a lot about the usefulness of the lunar method. A Board of Longitude was set up to examine the ideas offered by inventors and to reward the creators of promising ideas, so that they could continue to work on them. The Board consisted of learned gentlemen from the Royal Society, the Universities, the Admiralty and so on. Sir Isaac Newton was a member and he was convinced that the answer lay in the stars, moon and sun, as were others. This led to attempts to simplify the working of the lunar distance, or to catalogue the positions of the various heavenly bodies.

There was another and better method for finding the longitude. Give or take an occasional minor aberration, the world rotates 360º every twenty-four hours. If 360º of longitude are divided by twenty-four hours the answer is 15º per hour. In other words 15º of longitude could be thought of as being equivalent to one hour.

Naturally, it was not quite that simple, but a clock that could carry Greenwich time around the world would supply the basis of a mathematical equation that would give the navigator his longitude. Once a clock had been designed that could withstand the rigours of life afloat, the vexed question of establishing longitude at sea would be well on the way to being solved, and not by the lunar-distance solution. But there appeared to be little hope of that clock being developed, at least until John Harrison came on the scene. He was born in Yorkshire on 24th March 1693, the eldest of five children. His father was a skilled carpenter and, as soon as John was old enough, his father trained him in the skills of carpentry. Yet clocks were to be Harrison's passion.

John Harrison, carpenter, built his first clock before he was twenty and, what is more, he made it almost entirely of wood. Why he became interested in clocks, no one seems to know; he certainly never served an apprenticeship with a clockmaker. Whatever his motivation might have been, his wooden clock is now on show in the little museum maintained by the Worshipful Company of Clockmakers in the Guildhall, London. The wood for each part was carefully chosen: boxwood for the axles, oak for the cog wheels, but not any old oak. For the cog wheels Harrison chose oak, cut from a burr, the convoluted grain of which runs from the centre to the tip of each tooth.

He made two more similar clocks, one in 1715 and the other in 1717 and went on to build his reputation as a clockmaker. In 1720 Sir Charles Pelham had him build a clock for the tower of the stables on his estate at Brocklesby Manor. Harrison completed this clock in 1722 and, except for one brief stop, it has been running ever since.

By the summer of 1730 John Harrison went to London to put his plan to the Board of Longitude. It is a measure of the complexity of the problem that the response to the Board's offer, of what were at the time very large sums of money indeed, to someone who could come up with the means of solving longitude-finding at sea, had been poor, so poor, in fact, that the Board had not found it necessary so far to convene a meeting. Harrison knew that Dr Edmund Halley, the Astronomer Royal, was also a member of the Board, and he made his way to Greenwich to meet him.

Halley listened to Harrison and studied his drawings, realising that he was at last looking at an approach that had a chance of success. He also knew that the Board was dominated by academics, who were only interested in an astronomical answer to the problem and would not welcome a mechanical solution. Rather than expose the self-taught Harrison to what would have been an unsympathetic meeting with members of the Board, he sent him to see George Graham. Graham was a Fellow of the Royal Society and a first class watch- and instrument-maker. The two men spent a long day discussing clocks and watches, well into the evening, and Harrison's plans in particular. Graham was so impressed with Harrison that he invited him to dine and then sent him on his way with a generous interest-free loan to enable him to develop his ideas further.

Before a chronometer could be expected to function at sea, there were a number of basic problems to be overcome. Most of the clocks of the day depended for their movement on a weight suspended on a chain that ran over a cog wheel as the weight was lowered. The length of the pendulum regulated the speed of the movement. Neither the chain, the weight nor the pendulum would function properly, when subjected to the erratic movement of a ship at sea. Harrison replaced the traditional chain and weight by a spring, that had to be wound each day, and used an escapement in place of the pendulum. Springs had been used to power clocks and watches before, but when the spring was rewound each day, the movement of the clock either stopped or behaved erratically. Some makers had attempted to use a variety of lubricants to overcome this problem. Unfortunately, this led to further difficulties. The lubricants in use at the time deteriorated with age. They became more and more sticky, retarding the movement of the timepiece, which required frequent cleaning and meant stoppages. Harrison's answer was to make virtually friction-free bearings, that did not require a lubricant.

The other major problem was caused by the extremes of temperature in which the chronometers were required to work. The expansion and contraction caused by the temperature changes affected the rate at which the mechanisms worked. Harrison overcame this by using bimetallic strips of two metals with different coefficients of expansion. These were welded together so that the expansion or contraction of one strip of metal overcame or compensated for the other.

H-1, as John Harrison's first sea-going clock was named, took him and his brother James, five years to build. The finished timepiece (pictured opposite) was somewhat unwieldy: it weighed seventy-five pounds and was housed in a glazed wooden cabinet, measuring four feet square.

Opposite: *H-1, John Harrison's first sea-going clock*

John Harrison took his masterpiece to George Graham, as he had undertaken to do. Graham was so impressed with H-1 that he took it to the Royal Society and the members present shared his enthusiasm.

Under the terms of the Longitude Act, the Admiralty was required to check the performance of H-1 by sending it to the West Indies and back. The Admiralty allowed a year to go by, before arranging for sea trials and then, for reasons known only to themselves, sent it on a round trip to Lisbon, in HMS *Centurion*, not to the

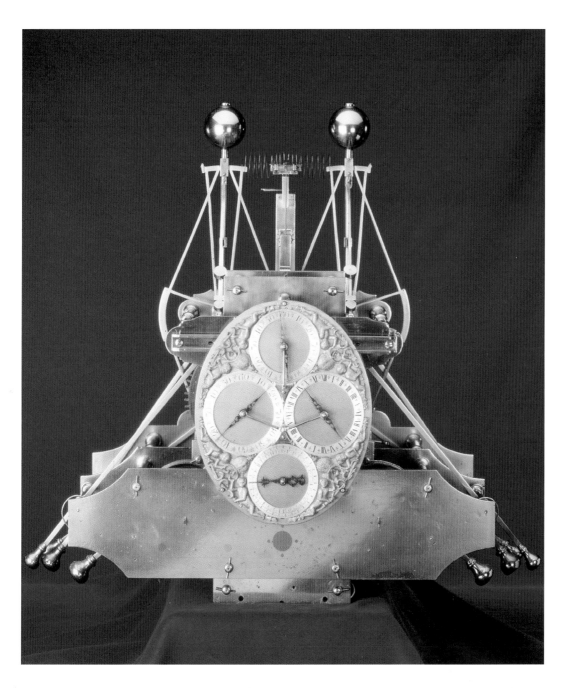

West Indies as the act stipulated. Captain Proctor had H-1 installed in his cabin, aboard *Centurion* and did all he could to help Harrison with the observations required. Sadly, Captain Proctor died shortly after reaching Lisbon and before he had a chance to write a report on the performance of H-1.

The clock and Harrison were transferred to HMS *Orford* for the home passage. The weather was bad and, instead of a week, it took Master Wills a month to make a landfall on England's south coast. His reckoning put him at Start Point, not far from Dartmouth. Harrison had also calculated their position and, with the aid of H-1, he claimed that they were off the Lizard, not far from Penzance, sixty miles from Wills's landfall. Further plotting, as they sailed along the South Coast, proved that Harrison's position was the correct one. Wills, in his report, admitted his own navigational error and praised, both H-1 and Harrison, most generously.

Eight members of the Board of Longitude met for the first time in twenty-three years to examine H-1. They were enthusiastic, but there was one dissenting voice: that of George Harrison's. Most people would have dwelt on the good points of their clock, but not Harrison, he drew the committee's attention to the faults in H-1 and told them that if they would grant him a further £500, he could eliminate those faults and make a smaller successor to H-1.

When it was finished, H-2 weighed in at eighty-six pounds but, as promised, it fitted into a smaller case. It had a number of innovations. This second timepiece did not go to sea, but was tested ashore rigorously. It was heated and cooled, was subjected to violent and erratic motion for several hours at a time and the Royal Society was well satisfied with the results of these tests. Harrison, on the other hand, was once again dissatisfied and told the Royal Society so.

There followed a period of nearly twenty years, which Harrison devoted to building H-3. The Board of Longitude saw him only when he requested a further £500.

In 1753, John Jefferys, a Freeman of The Worshipful Company of Clockmakers, made a pocket watch as a gift for Harrison. Jefferys had built some of Harrison's innovations into it and Harrison was delighted with the timepiece. The pocket watch inspired him to miniaturize H-4. When it was completed in 1759, it had a diameter of five inches and weighed three pounds. As a time keeper, it was near faultless. In the interval between the completion of H-2 and of the watch, H-4 , changes had

Opposite: *H-4, John Harrison's watch built in 1759.*

taken place. New star tables and other sources of information were now appearing, along with some simplification of the observation and of the mathematical workings of the lunar distance. There were new faces on the Board of Longitude and at the Royal Society. James Bradley, the third Astronomer Royal, was a close friend of Nevil Maskelyne. Both men were heavily involved in promoting the lunar-distance solution and had an eye on the £20,000 prize. They were in the process of compiling tables that gave the position of stars and other heavenly bodies, suitable for navigation A prize of £20,000 was worth fighting for, and fight they did, using every dirty trick in the book. With members of the Board promoting their own interests, it was not the even-handed organisation it should have been.

At last, H-4 was given its chance. William Harrison, a son of John Harrison, was to take the watch aboard HMS *Deptford* to take passage to Port Royal, Jamaica. The weather was bad and it took them eighty-one days for a passage that normally lasted three weeks. H-4 had lost an unbelievable five seconds in eighty-one days. During our own voyage, we used a quartz-crystal watch for our celestial sights and, in eighty days, it lost twenty seconds. The loss was consistent and caused us no problem, but it was designed 200 years after Harrison's watch was developed!

Harrison was a self-taught man of humble background and, in spite of his immense achievements, he was regarded as an artisan by some of the members of the Board of Longitude and of the Royal Society, who resented the fact that the prize money should go to someone who was not one of their peers. They managed to deny him his full reward for forty years, but Harrison was finally recompensed in 1773, but it required no less than the intervention of King George III!

Meanwhile, the instrument-makers were creating their own revolution. The astrolabe, the cross staff and back staff were being transformed into the modern sextant, that is truly a precision instrument. Ironically, all this effort would be made redundant relatively quickly by the invention of a little black box, the size of a mobile phone, stuffed with electronics. The GPS (ground position satellite) will pick up signals from satellites that are orbiting the earth and give a position, correct to within a metre or two, which is amazing enough, but the military version of this instrument has an accuracy that is measured in centimetres.

By mid-April 1772, *Resolution* and *Adventure* were very nearly ready to sail. Every inch of space aboard *Resolution* was in use, to the point that there was not room enough for the seamen's traditional wooden sea chests and Cook replaced them by canvas bags. The last of the stores were taken aboard on 25th of April. Cook was concerned about the seaworthiness and the manoeuvrability of his ship. Even when she was moored to the dock wall at Woolwich, she felt wrong. It is impossible to explain, but when you have lived aboard sailing vessels for a number of years, they either feel right or they don't. It is not something that can be put into words, but most people develop this sixth sense and, in a man with Cook's background and experience, it would have been highly developed.

Banks and his party came aboard on 2nd of May and celebrated their arrival by giving a party for the Earl of Sandwich, the French ambassador and sundry notables. A small orchestra played for the entertainment of the guests, who were waited upon

by servants dressed in scarlet and silver livery. When the party was over, Banks and the other gentlemen left to make their way to Plymouth, where they would await the arrival of *Resolution*. Cook also left the ship to attend to some personal business, intending to rejoin her in the Downs.

Cooper, the First Lieutenant, was placed in charge, with orders to take the vessel from Woolwich to Dover, a day's sail away if the wind was right, and there to await Cook's arrival. Cooper took a pilot on board to see *Resolution* safely through the shoals that abound in the Thames Estuary and on round to Dover. After four days they had failed to reach their destination and the pilot refused to continue with the passage. He complained that the ship was not seaworthy and would take her no further. *Resolution* was indeed cranky, she did not answer to her helm as she should. Even in the lightest of breezes, she heeled badly, when other vessels sailed upright. Banks' friends themselves were horrified. Cook's instinct had been correct. The alterations had turned a responsive vessel into 'an exceedingly dangerous and unsafe ship.'

To raise the decks, as Banks had demanded, entailed the addition of a great deal of extra timber. If the deck was raised a foot, it would require the sides of the vessel to be raised by the same amount. Added to this, was the not inconsiderable weight of the new cabin on the upper deck. Doubtless, the increased space would have been used, at least in part, for stowage. All this extra weight was placed above what was originally the highest point of the hull. The result was to raise the centre of gravity of the vessel, well above its original position, robbing her of her stability.

On his return, Cook was given a detailed report on *Resolution*'s problems. The Navy Board took immediate steps: *Resolution* was ordered to Sheerness, where the infamous modifications were to be removed. She was once more invaded by gangs of workmen and, again, time passed. Cook blamed himself for not trusting his instinct and for not being firmer, when Banks had issued his ultimatum.

When Banks learned what was happening, he protested fiercely and threatened to leave the expedition. It seems that the threats were made in the hope of convincing Cook and the Admiralty that *Resolution* should be abandoned and a 'more suitable

Opposite: An eighteenth-century sextant, similar to the ones used by James Cook aboard Endeavour. This instrument was used to measure angles and takes its name from its curved edge, which is an arc one sixth of a circle. The map under the sextant is reputed to be the one shown under Cook's hand in the famous portrait by Nathaniel Dance, dated 1776, and reproduced on the jacket of this book.

vessel' acquired. Banks wrote letters of complaint to anyone he thought might help his cause. He claimed that the ship would still be unseaworthy, even if she was returned to her original design, and that the changes would be detrimental to the health of the crew, who would have less space... and so on. It was obvious that the welter of complaints did not stem from a concern for the health of the crew, or anyone else's, but for his own personal comfort and aggrandizement. He was also enraged at being proved wrong. In an effort to pacify him, it was pointed out to Banks, that even after *Resolution* had been returned to her original state, his accommodation would be considerably larger than aboard *Endeavour*. What is more, the Great Cabin would give him at least as much room as an Admiral of the fleet could expect to have.

What antagonised the Navy Board in particular, was Banks's insistence that he should have been consulted on the choice of vessel. The two points of view could never be reconciled. Banks believed that his work was the prime purpose of the voyage, while the Navy looked upon him as a rather troublesome passenger, with a useful, but lesser role to play.

When Banks saw *Resolution* after she had been restored to her original design, he exploded and immediately ordered that all his gear and servants be put ashore. Banks was to remain convinced that the Navy Board had deliberately altered *Resolution* in a manner calculated to make her unseaworthy, in order to get rid of him.

Naturally, the friendship between Cook and Banks cooled somewhat in the aftermath of this disagreement. Later, when Cook reached Cape Town, he wrote a conciliatory letter to Banks who replied in a similar vein and the fences were mended between these two men, who had achieved so much together. It is good to be able to state that Banks, in later years, would remain a firm supporter of Cook and his work.

Banks, with Solander and the others did not languish ashore for long, they joined an expedition that was to carry out some exploratory work in Iceland. Although Banks had been so difficult, at times even to the point of arrogance, he had shown that it was rewarding to carry one or more naturalists of his calibre on the sort of expedition Cook was undertaking. Accordingly, and at very short notice, replacements were sought for *Resolution*'s voyage.

Inquiries were made and John Reinhold Forster's name came up. He was of English descent, but born in Prussia, and had recently returned to England with his family. Forster was a clergyman, but was much more interested in science, botany, zoology and languages, than he was in religion. He hoped that he would be able to find work as a teacher, which he did, but he was far from suited to that profession. In less that two years he was dismissed from two teaching posts. In late 1770 he moved to London and worked as a translator, where amongst other tasks, he translated Bougainville's account of his circumnavigation into English.

In May 1772 the Admiralty asked Forster if he would care to take Bank's place aboard *Resolution*. Forster accepted, but put forward two conditions. The first was that his eighteen-year-old son, George, should be allowed to sail with him. This presented no problem, as the latter was a well-trained naturalist and a skilful draughtsman. The second condition concerned the financial support of the rest of his family, while he was away. This hurdle was also overcome and, on 20th June, the Forsters got their gear aboard *Resolution*, which was anchored in the Thames, undergoing some last-minute repairs.

Cook was months behind on his schedule and he must have begun to think that the expedition would never get underway.

Dolphin

12. England to New Zealand

July 1772 to April 1773

Overnight, the wind had changed to a north-westerly, which allowed Cook to leave Plymouth and, at 0600 on 13th July 1772, *Resolution* and *Adventure* set sail and stood to the south-west. Their first stop was to be Madeira, where they intended to replenish their stock of fresh food, top up their water supplies and purchase a quantity of wine.

The winds experienced by the two vessels on passage to Madeira were often strong, but mostly in the right direction, and that is what counts. By the end of the month they were anchored off Funchal. Their business was soon concluded and they were back at sea on 2nd of August. Eleven days later both ships were anchored at Porto Prayo, in the Cape Verde Islands, and the never-ending quest for fresh food and water started again.

Supplies were not as plentiful as they had been in Madeira, but there were some and they were quickly bought up. There were only limited quantities of water and wood was almost non-existent. There was no reason for them to linger and at 2100 on 15th of August, *Resolution* and *Adventure* were at sea again, their destination Cape Town.

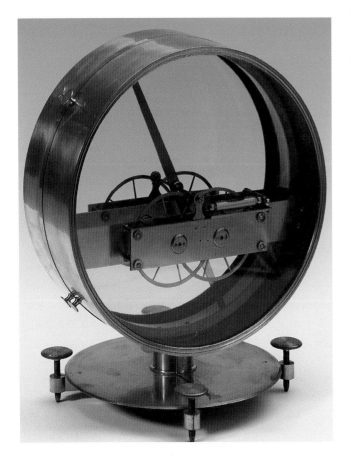

Their passage to the Cape was a slow one. Part of this leg took them across the Doldrums, where they spent much of their time becalmed. Even when the wind blew, it often did not serve. Daily runs of less than 100 miles were commonplace. Had the wind been favourable, the two vessels could have expected to make daily runs of 120 to 150 miles. But nothing lasts for ever, not even a protracted passage. Early in the morning of 30th August, both ships dropped their anchors in Table Bay, where they were to stay for almost a month. Those crew members not engaged in revictualling their ships, worked hard to get the vessels into tip-top condition, in preparation for the tough time ahead.

Wales and Bayly, the astronomers, went ashore to test the chronometer made by Larcum Kendall – an exact replica of Harrison's design – along with three other timepieces. The Kendall model had been tested by Nevil Maskelyne on his voyage to Barbados, and he had to concede that it was remarkably accurate. Because the longitude of Cape Town was well established, it would allow the behaviour of the chronometers to be tested reliably against astronomical time.

The Forsters worked hard at their tasks, but Banks's open-handed generosity on the earlier voyage meant that the local population had greater expectations than Forster could fulfil by way of payment for specimens of the local flora and fauna. To offset these difficulties, Forster and his son had the good fortune to meet a young Swedish naturalist, Anders Sparrman, who had been trained by Linnaeus, no less. They were so impressed by the young Swede's learning and ability that they persuaded Captain Cook to allow him to join them.

The Dutch Governor of South Africa remembered Cook from his previous visit and welcomed him warmly. He had news of two French ships, seeking *Terra Australis*, only they referred to it as *France Australe*. Yves Kerguelen was leading the French expedition. About eight months earlier, he had discovered land to the south of Mauritius, in latitude 48º south. He claimed, without further evidence, that this land was the central part of the coastline of *France Australe*. In fact, what he had discovered was an island that now bears his name. In 1503 a countryman of his, Gonneville, claimed to have found the landmass, *he* named *France Australe*. This elusive continent would go on exercising its attraction on French explorers.

In 1739, Jean-Baptiste Bouvet de Lozier was on that very same quest. In his day, the belief that seawater could not freeze, still persisted. Through thick icy fog he could see snow and ice and assumed he was approaching land. His belief was reinforced when he saw icy cliffs looming in the distance. Could this be the fabled land glimpsed by de Gonneville? He named it Cape Circumcision, after the feast day on which he discovered it, ie 1st of January. Foul weather made it impossible for Bouvet de Lozier to examine his find further, but that did not stop him from claiming that he had found *France Australe*. He added this small piece of the puzzle to his chart, and Cook was determined to find his way to it.

Opposite: *Dip circle made by Edward Nairne in 1772. Commissioned by the Board of Longitude, this instrument was used to measure the vertical component of the earth's magnetic field. It is reputed to have been used by Cook on one or more of his voyages.*

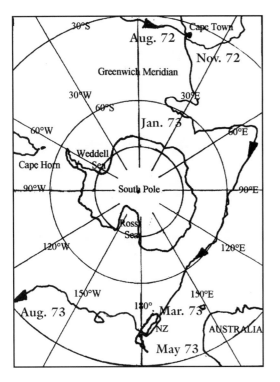

Cook's orders required him to complete the circumnavigation of Antarctica in an effort to settle, once and for all, the vexed question of *Terra Australis*. To achieve this, he first had to head south, until he found the disputed landmass, or could safely say that no such continent existed. If he found land, he was to lay claim to it for the King, explore it and chart the coastline, learn all he could of its people and of its flora and fauna. If nothing was found, Cook was to continue eastwards, still searching for the Great Southern Continent, until he was back at Cape Town.

On 23rd November 1772, at 1500, *Resolution* and *Adventure* raised their anchors and set sail. Once they were on their way fifteen-gun-salutes were exchanged with the garrison. The wind was blowing hard, and they had to make several tacks to get themselves clear of Table Bay. Once out, the course was altered to the south-west to gain a safe distance offshore before turning south.

Cook had decided to search for Bouvet de Lozier's discovery, although by now he was highly sceptical of the tales of a Great Southern Continent, but Bouvet de Lozier's sighting was as good a starting point as any, and he headed south in search of it.

Cook spent some time cruising in the region of the reported position of Bouvet's discovery, but failed to find any sign of it. What he did not know was that the longitude given by Bouvet, was seven degrees in error, a substantial distance. After several forays, Cook seriously doubted that such a place existed. What is remarkable is that the Frenchman had come across tiny Bouvet Island at all. This tiny speck of land lies over a thousand mile in any direction from the nearest shore. Its exact position was not plotted until 1898.

The weather was mixed, ranging from gales to calms, but on the 30th it was blowing so hard that for twenty-four hours both vessels lay to, with just a mizzen staysail raised, to keep the ships' heads up to windward. Things were so bad that the cooking fires could not be kept burning with any degree of safety. Both ships'

Opposite: This engraving, based on a drawing made during the voyage, captures the total allienness of the landscape Cook and his men found themselves amongst.

crews were cold and wet and deprived of sleep by the violent motion. To add to their discomfort, *Resolution*'s deck seams leaked badly, despite the shipwrights' best efforts to caulk them.

On December 10th, they met their first icebergs. From now on, ice was never to be out of sight from the two vessels, until after they turned to the north and headed east in the direction of New Zealand.

Falling rain, snow and sleet froze, wherever it settled, covering rigging, sails and deck in ice, with huge icicles hanging from the rigging and the yards, a potential danger to the men working beneath them. The duty watch spent much of their time breaking up the ice and casting it overboard. Had it been allowed to accumulate, the vessel would have become top heavy and, therefore, in danger of capsizing.

Cook had issued cold weather gear to all the men – the famous 'fearnoughts' – but it was not particularly good at retaining the body's heat and incapable of keeping the wet out for very long. The conditions must have been dreadful and yet it seems that the crew went about their daily routine without complaint. I suspect that the Navy's unofficial motto: 'if you can't take a joke, you shouldn't have joined,' was brought out for an airing on these occasions.

Both vessels had a number of pigs, sheep and poultry on board and the cold weather was killing them off at a steady rate. Whenever a sufficient number of the poor beasts had succumbed to the low temperatures, the ships' companies were served fresh meat.

It's an ill wind that blows nobody any good!

It was not unusual for thirty or forty icebergs to be in view from the ships at any one time. Frequently, rain, sleet, snow and thick fog reduced visibility so badly as to make sailing among the many icebergs hazardous in the extreme. Most nights would find both vessels under much reduced sail.

Cook continued to press on southwards, encouraged by the large number of birds and seals visible to them. The current belief was that these creatures would only be found close to land. By and large it is true, but there are exceptions. During our own journeys we encountered small birds, skimming the surface of the sea in mid-ocean. When we took a closer look, we found that they were catching the insects that live on the surface of the sea.

By the middle of January, the two vessels had penetrated as far south as 64.5º of latitude. As they worked their way south, the ice became more and more densely packed, until it was obvious to them that it would be foolhardy to continue southwards. Cook headed north to escape the ice, turning to the east whenever the wind and ice allowed him to do so, the search for the Southern Continent always dominating his thinking.

By now both vessels were running short of water and it was decided to collect ice from icebergs nearby and melt it in the hope that it would not be too salty. Parties were sent out to cut blocks of ice and, to their delight, found that when melted, the ice yielded fresh water, without a trace of salt. On their first attempt, Cook's men gathered fifteen tuns of water.

Cook discovered that two of his crew had been aboard ships that had been caught in the ice, off Greenland, one for six weeks and the other for nine weeks. Typically, he sought them out and learned what he could from them.

The expedition had sailed about 200 miles to the east and Cook decided to continue in that direction, along the edge of the ice, turning south whenever the opportunity arose. All the time the temperature hovered around freezing point and, on 20th December, Cook set all the tailors to work, lengthening the sleeves of the crew's jackets and making caps to help keep the men warm.

The weather was mixed, the wind ranging from near calm to gale force. Visibility was often badly reduced by rain, sleet, snow or by fog. When this happened, the two vessels moved up and stayed close together to avoid being separated. Still, icebergs were everywhere and, in the poor visibility, they were a constant danger.

In spite of the difficult conditions, the scientific side of the voyage was not neglected. Specimens of bird life were shot and brought aboard to be studied by the scientists and, whenever the sky cleared, sights were taken to keep track of their position. When the wind dropped, the direction and strength of the currents in the sea were noted. The magnetic variation was also calculated regularly.

Cook stayed as far south as the ice would allow, most of the time in the low-sixties latitudes, making slow progress to the east, whenever possible.

At about 0830 on 8th of February, *Adventure* was sailing about a mile or so, astern of *Resolution* on her port quarter, when thick fog closed in quite suddenly. Both vessels lost sight of each other. Half-an-hour after the appearance of the fog, Cook had a gun fired, expecting *Adventure* to respond, but there was no answering gun shot. It seemed impossible that within such a short time, the two ships should have sailed far enough away from each other, as to be unable to hear the gun.

A procedure had been agreed upon, should such a situation arise: fires were to be displayed at the main masthead, constant watch kept and guns fired at pre-arranged intervals. The ships were to beat back to the position at which they had last been together and cruise for three days around that area. If they still did not meet up, they were to make their separate ways to the rendezvous in New Zealand, continuing with the scientific work en route.

Resolution fired a gun, every hour, until 1400. As the hourly gunfire failed to evoke a response from *Adventure*, the period between discharges was reduced to every thirty minutes and, after dark, fires were displayed at the masthead. In the meantime, Cook had sailed a series of courses, which he hoped would bring him closer to *Adventure*, but all to no avail. The fog stayed with them, thinning from time to time, only to thicken again. As if that was not enough, the wind picked up, requiring them to shorten sail.

Wednesday 10th brought some improvement in the weather. The gale abated and visibility improved until the masthead lookouts could see twelve to fifteen miles in all directions, but there was still no sign of *Adventure*. By late afternoon Cook decided that it was a waste of time, hoping to meet up with her. The constant lookout and the hourly gunshot were discontinued, to be replaced by the usual half-hourly check from the masthead.

Resolution resumed her easterly progress towards New Zealand, staying in the Antarctic latitudes and enduring foul and freezing weather. On Thursday, 25th of March, the weather finally improved and, by 1000, the lookout sighted the southern part of New Zealand.

They were not to get into shelter until the following day. It was dark when *Resolution* reached land and it was necessary for the ship to stand off and on for the rest of that night, rather than attempting to enter an intricate waterway in darkness.

At first light on Friday morning, they turned towards the land and entered Dusky Sound at about noon. They found themselves a suitable anchorage in fifty fathoms of water, close to the shore.

Resolution had been at sea for 117 days, without sight of land. Cook and his men had covered almost 11,000 nautical miles, much of it in freezing temperatures, yet out of a crew of 120, only one man was down with scurvy. For those days, it was a remarkable achievement!

Queen Charlotte's Sound was the agreed rendezvous, but Cook was in no hurry to meet *Adventure*. He had not called in at Dusky Sound, on his earlier visit to New Zealand, and he could never pass up a chance of exploring new territory.

As always, obtaining sufficient quantities of fresh food for the ship's company was to the forefront of Cook's mind. A boat was sent out to fish and it came back with as much as all hands could eat for their supper. Another boat was taken to a nearby rock on which seals were sunning themselves. The men killed one and took it back to the ship to provide fresh meat. The next morning the boat was out fishing again to supply fresh fish for the midday meal.

The anchorage they had chosen for their overnight stay was less than perfect. Cook went to the northwest side of the bay in one of the ship's boats and Mr Pickersgill was sent to the southeast, both on the lookout for a better anchorage. Cook found a spot, but when he spoke with Mr Pickersgill, it was obvious that the latter had found a far more suitable place.

Opposite: Chart of New Zealand *by James Cook*

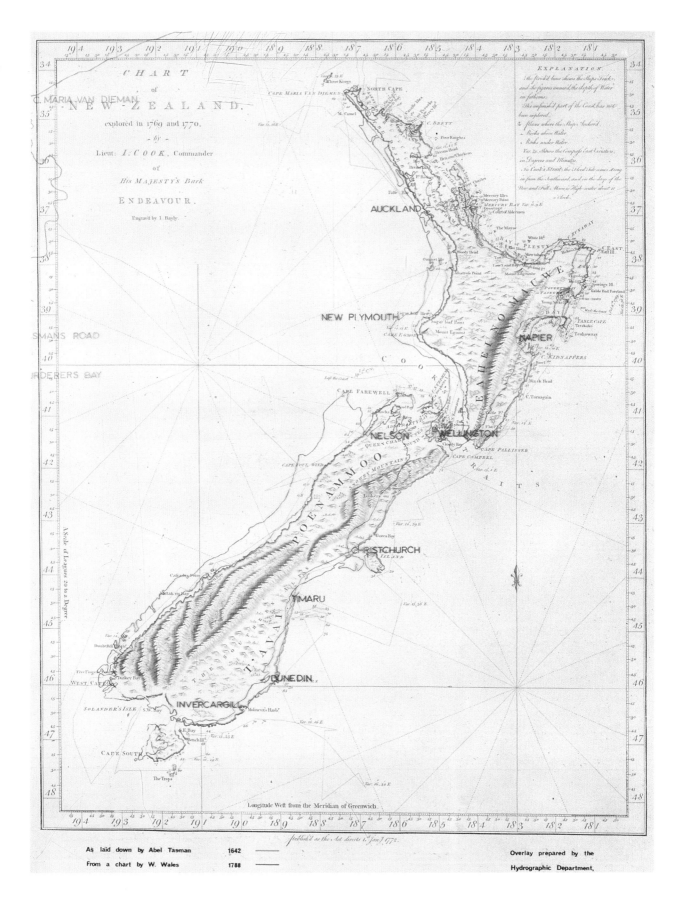

At about 0900 sail was made, the anchor brought aboard and, with a gentle breeze blowing in the right direction, *Resolution* sailed into what is now known as Pickersgill's Harbour, which proved to be very snug.

Resolution was moored head and stern to trees, a large branch of one serving as a temporary brow to enable them to step ashore, dryshod. A good supply of fresh water and wood was close to hand. The bay seemed to have a never ending supply of seals and fish and, ashore, there were wild fowl to be trapped or shot. After their long period at sea, they must have felt that they had reached paradise.

The scientists were put ashore and working parties were set to collecting fuel and water while others sought fresh food. They also gathered quantities of pine needles, from trees that resembled spruce, to make a form of beer by steeping the needles in warm water, with a small amount of yeast. Cook had noted that the inhabitants of Newfoundland made a similar concoction, which they drank as an anti-scorbutic throughout the winter, when vegetables were not available.

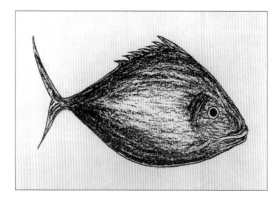

Captain Cook encouraged the crew to catch fresh fish as it was a valuable addition to their diet.

13. Pickersgill Harbour to Tahiti

April to September 1773

By 1st of April 1773 work parties were ashore, clearing space and setting up a variety of workshops. The sailmaker had plenty of work to do, repairing damaged sails and making new ones, to replace those that were too badly damaged. The forge was set up and the coopers busied themselves repairing casks, before others filled them. The astronomers selected a suitable place for their observatory and the botanists worked their way inland, in search of new specimens for their collections. Cook and his officers were engaged in the inevitable survey work of the coast, but also found time to explore further afield.

A week passed before there was any contact with the local population and, even then, it was rather limited. Initially, the locals were cautious rather than hostile, but they slowly gained in confidence. About ten days following the first timorous meeting, they were confident enough to set up their own camp, about a hundred yards from *Resolution*.

Cook stayed in Dusky Sound until 9th of May and, reading the accounts of their stay, it must have been a very pleasant one. A great deal of work had to be done in preparation for the next stage of their voyage, but it was done in congenial surroundings, with weather that must have been a vast improvement on the conditions they had endured in the Antarctic.

By 12th of May *Resolution* was well on her way to the rendezvous in Queen Charlotte's Sound, where Cook hoped to be reunited with *Adventure*. The weather for this leg was mixed: gales followed by calms, the former ranging from what Cook described as 'strong gales to gentle gales.' The only thing of note in this passage was that *Resolution* sailed for the best part of two days, in the company of numerous waterspouts.

Resolution entered Queen Charlotte's Sound on 19th May, at about 1800, and anchored within easy distance of *Adventure*. Cook learned from Captain Furneaux that, when the fog closed in on the two vessels in the Antarctic, *Adventure* had been about three miles from *Resolution*. The fog appeared so suddenly and was so thick that, within seconds, both vessels were lost to each other. Some of the crew of *Adventure* heard *Resolution*'s first shot and replied to it at once. *Adventure* fired a shot every half hour for two more days, but no reply was heard. In all probability, the dense fog muffled the sound of the shots.

Captain Furneaux turned to the west and made his way towards New Zealand. Like Cook, he recorded everything of note en route. But unlike Cook, Furneaux went round the south coast, and then northwards along the eastern coast of Tasmania, before turning towards New Zealand. As he sailed along these coasts and before leaving for Queen Charlotte's Sound, he managed to record details of about two thirds of the coastline. The information he compiled led Cook to believe that Tasmania probably formed part of the Australian mainland.

It had been Furneaux's intention to spend the winter in Queen Charlotte's Sound, but Cook's arrival put paid to that idea. Cook had decided to spend the winter months, looking for land to the east of New Zealand, as far as longitude 140º west and within the latitudes of 41º and 46º south. The latitudes he had chosen would allow them to run before the westerly winds, until they approached the longitude of Tahiti, then the course would be altered to the north, until they reached that island. After refitting and revictualling, the two vessels were to return to New Zealand by the most direct route. They would stock up wood, water and what fresh food they could gather, before proceeding to the south and exploring the area between New Zealand and Cape Horn.

Furneaux had never really bothered with Cook's anti-scurvy measures and, as a result, the health of his crew suffered. Whilst the two barques were anchored in Queen Charlotte's Sound, Cook himself went ashore in search of anti-scorbutic food and, when it was found and collected, he ordered that both crews were to have these items included in their diet. One sailor wrote: 'it was the custom of our crews to eat almost every herb plant root and kinds of fruit they could possibly light upon.' Those men who had sailed in other ships, would have known that the absence of disease aboard Cook's vessel was quite unusual and, they also quickly learned, that falling in with Cook's views, could earn them brownie points. The same sailor wrote that they all 'knew that it was a great recommendation to be seen coming on board from a pleasure jaunt with a handkerchief full of greens.' While both ships were in New Zealand, the health of the crew of *Adventure* improved but, as soon as they separated, and Cook was no longer there to enforce his instructions, Furneaux once again neglected the anti-scorbutic measures.

Opposite: William Hodges's dramatic representation of Resolution, *off Cape Stephens, and surrounded by towering waterspouts.*

Cook intended for both ships to spend the coming summer in the Antarctic, sailing between New Zealand and Cape Horn. Tahiti would be their rendezvous point, should they become separated again.

At about this time, the winding mechanism on one of the chronometers made by Arnold, failed to function, and there was no way that it could be restarted. This was the second of Arnold's watches to have stopped; the first having worked only as far as Cape Town.

By the beginning of June the two barques were ready to move on, and both vessels set sail on 7th. Once they were out of the Sound, the wind was blowing from dead ahead. No ship of that period could have sailed as close to the wind as a modern yacht can and they had to tack, ie proceed in a zigzag pattern, all the way through the Straits and it was not till thirty-six hours later that they were finally clear of the land. Cook and Furneaux took their departure from Cape Palliser on the south-eastern corner of North Island.

At the end of July, Cook had a boat take him across to *Adventure* to check on the health of her crew. He found the cook dead and twenty more men down with advanced scurvy and dysentery, while many others showed the early symptoms of

scurvy. On the other hand, *Resolution* had only one man suffering with scurvy. Clearly Captain Furneaux had not bothered to enforce the Commander's instructions on the inclusion of anti-scorbutics in his men's diet and they were 'much inflicted' as a result. Furneaux was endangering, not only the crew's health, but his behaviour could well compromise the long-term success of the voyage. Cook sent a man from *Resolution* to replace the dead cook, expressing his disappointment to Captain Furneaux and giving him *written* instructions on how to handle the problem. He also confided his displeasure to his journal.

By now, both vessels were heading to the north, with Tahiti as their target. Having failed to find land this far, Cook was more convinced than ever that there was no *Terra Australis Incognita* to be found in the Pacific, but to convince the doubters back home, he had to sail through the remaining areas.

On 8th of August they found the trade winds and the pleasant weather that goes with them. Three days later, at 0600, they sighted the first land since leaving New Zealand. It was one of the small atolls that make up the Tuamotu Archipelago. Cook did not linger, as he was quite concerned about the state of the health of the crew of *Adventure* and wanted to get to Tahiti, where he was sure he could bring them back to health, as fresh food would be available.

The string of atolls that lay in the path to Tahiti made it necessary to shorten sail every night and Cook chafed at the delay. On 15th August he summoned Furneaux aboard *Resolution* and told him that he intended to put into the nearest point on Tahiti for the sake of those who urgently needed fresh food to recover their health.

This was Oaiti'peha Bay, on the south-east corner of the main island. By midnight it was clear that, if they continued under sail, they would reach the bay while it was still cloaked in darkness. Conditions were such that they could heave to and wait for the sun to rise. By 0400 it was light enough to make their approach to Oaiti'peha Bay.

Cook had gone below to snatch a few hours' sleep before entering the bay. He had left orders for the course to be steered, when they resumed sailing. When he awoke and went on deck, he found, to his dismay, that the course being steered was wrong and that *Resolution* and *Adventure* were only a mile or so from the fringing reef, and steadily getting closer. A mile to the reef! A vessel making five knots would

Opposite: HMS Resolution and Adventure in Matavai Bay, *Tahiti*, *painted by William Hodges*, ca *1773*.

require some twelve minutes to cover one mile and they would need to do a great deal of work in order to make a substantial alteration to their course.

No sooner was *Resolution* turned to her new course that it was the cue for the wind to drop. Boats were lowered to tow *Resolution* and *Adventure* clear, but they were unable to stop the two vessels from slowly drifting towards the reef. An opening appeared in the reef, Cook hoped to sail through it into the lagoon, and anchor, but some of the Tahitians who were on board told Cook that it was too shallow. A boat that had been sent to check the depths in the lagoon signalled the same message. As the tide rose, the sea raced through the narrow opening in an effort to fill the lagoon, and it was this indraught that was dragging the two ships towards the reef, despite the best efforts of the men in the pulling boats.

As soon as the water shoaled sufficiently, both vessels dropped an anchor which stopped any further dragging, but by then, *Resolution* was so close to the reef that, although her anchor held, with every rise and fall of the sea she hammered her keel so hard that those on deck had difficulty in staying on their feet.

Adventure's anchor held but her stern was so close to *Resolution*'s starboard bow that they were in constant danger of colliding. *Resolution*'s boats laid out two kedge anchors and when these were hauled on she began her slow move to safety. Late in the afternoon a light breeze came in and, with its help, *Resolution* was moved some two miles offshore. With *Resolution* safe, Cook was able to send the boats to help *Adventure* but, before they could reach her, she was moving under the influence of the breeze. *Adventure* soon joined Cook and the night was spent making short boards on and off the coast and, at dawn, they made their way into Oaiti'peha Bay. The sick men from *Adventure* were put ashore each day and, with a diet of fresh food, they were soon on the mend. The ships stayed for nine days and then left for Matavai Bay, reaching there late in the afternoon of 26th August.

The incident at Oaiti'peha Bay was of particular importance because it pointed to a definite change in Cook's character and behaviour. His annoyance at the helmsman's error, though understandable, expressed itself in a way which was a departure from his former mild behaviour. Anders Sparrman, the Swedish botanist on board, reported how the Captain, 'whilst the danger lasted, stamped about on deck and grew hoarse with shouting.' The crew would grow wary of these rages which they called 'heivas' – the violent stamping reminding them of a Tahitian dance of the same name. They looked upon them with amused affection at the beginning, but soon grew to fear Cook's terrifying mood-swings and violent anger. A man who had always impressed his officers and the men by his calm in a crisis, seemed now to be seething with tension. The men were confused, their beloved Captain had changed and it is Mr Sparrman who, again, gives an indication of the possible source of the problem.

He describes how, once the danger was past, he went down to the ward room in the company of Captain Cook, who was by then 'suffering so greatly from his stomach that he was in a great sweat and could scarcely stand.' Mr Sparrman dosed Cook with a large measure of brandy, which he described as a traditional Swedish remedy, and Cook seemed to improve. Interestingly, Cook does not mention his indisposition in his journal and it is open to conjecture whether it was a real medical problem or a phenomenon brought on by stress. It is however, borne out by the personal journals of his officers and of the scientists that the staggering rages, coinciding with bouts of severe pain, became a recurring problem during the voyage.

Opposite: Flogging was a frequent punishment aboard Royal Navy vessels. The culprit was tied to a grating and given a pre-ordered number of lashes, with the ship's company looking on.

14. Tahiti to Queen Charlotte's Sound

August 1773 to November 1773

Long before they were anchored, canoes full of Tahitians wanting to trade fruit and vegetables for axes and iron spikes, surrounded *Resolution* and *Adventure*. Most of the inhabitants knew Cook and he remembered many of them from his previous visit.

Cook made his way ashore to meet the various dignitaries who had arrived to greet him and, while this was happening, the ships' companies were busy erecting shelters and workshops on the sites they had occupied before.

They had hoped that they would be able to buy a number of hogs which, previously, were to be found everywhere but, this time, there were very few to be seen and even fewer to be purchased. There were probably two reasons for the shortage. One was that ships calling at Tahiti bought considerable provisions and the other was that the two major kingdoms had been locked in civil war, some time before. These factors may well have unbalanced the economy of the island.

Late in the evening of 30th August there was a loud commotion ashore and it was obvious that men from the ships were involved. An officer with an armed party was sent to investigate. Meanwhile Cook had the two ships' companies checked for missing men. Quite soon the armed party returned with four men from the shore party and a number of hands from *Adventure* too.

The full story never emerged, none of the men would admit to misdeeds of any kind, but it was believed that attempts to secure sexual favours from some of the local women were the reason for the riot. Cook, always anxious to avoid friction with the local inhabitants, put the culprits in irons for the night and the next day they were punished, some receiving a dozen lashes and others eighteen. It must have been a fairly serious matter, as the local population fled the area and did not return before the ships left.

By now the sick were restored to health, all the outstanding work had been completed and they had bought as much food as was available: it was time to move on. The tents were dismantled and brought aboard and all was made ready for sea.

From Matavai Bay a northwesterly course took the two vessels towards the northern end of Huahine. By dusk, on Friday, 2nd September, they were within twelve miles of Owharre Lagoon, on the western side of the atoll, where they hoped to anchor. All night they stood off and on until sunrise when *Resolution* cautiously made her way into the lagoon and anchored. *Adventure* was not so lucky, she got herself hung up in the wind, as she tried to tack, and grounded herself on the reef that lines the northern shore of the entrance to the lagoon.

Knowing there was no room for error in the pass that let them into the lagoon, Cook sent *Resolution*'s launch to go to the aid of *Adventure*. With the help of the launch, *Adventure* was soon off the reef and anchored safely, the only damage being to Captain Furneaux's pride!

Both ships were soon surrounded by the usual swarm of canoes, filled with locals anxious to trade fresh food, including a plentiful supply of hogs, in return for hatchets, nails and beads.

One of the first tasks for Cook was to pay his respects to the Chief, whom he had first met in 1769, renew his friendship with him and exchange gifts. Things went well and a happy rapport was renewed.

Sometime in the forenoon of Monday, Mr Sparrman was attacked and robbed by two men whilst he was out alone, collecting botanical specimens. They stripped him of everything but his trousers. Cook reported the matter to the Chief who was angry and most apologetic to Cook. He personally set off in pursuit of the two men who had robbed Sparrman. It took a while, but, eventually, the more important of the stolen items were recovered before the ships left.

As they were to sail the next day, Cook was anxious to restore good relations before they left. To underline his desire to leave on a happy note, he and his officers organised a reception on board *Resolution* and entertained the Chief and some of his family and friends. The usual exchange of gifts also took place.

Early on Tuesday, Cook, Furneaux and Forster called on the Chief to thank him for his hospitality and to bid him farewell. Meanwhile, the anchor cables were shortened and all was made ready to sail, as soon as the party returned aboard.

Daylight on Wednesday, 8th September, found the two barques off the entrance of Ohamaneno Lagoon on the western side of Huahine. The wind was blowing straight out from the pass into the lagoon. There seemed to be no hope of making way against the head wind in the narrow pass that would not give them space to tack. But Cook was equal to the situation, in his log he describes how 'we made a few short boards and borrowed on in.' To translate: the few short boards would be a few short tacks against the wind to build up speed and get close to the entrance of the pass. Once in position, he 'would borrow on in,' ie *Resolution* would be turned quickly into the wind and shoot through the pass, carried by her momentum. Once in the lagoon he payed off her head and used the wind again to take him to the spot in which he had chosen to anchor. *Adventure* followed *Resolution*'s example and anchored close by. A real seaman's gambit, requiring good judgment and a steady nerve.

As Huahine offered plentiful provisions, they stayed until 17th September and, when they left, they had as much food as they could possibly stow.

Resolution and *Adventure* sailed westwards with a little south in their course. They were on their way to New Zealand to refit before they tackled the Antarctic again, but it was Cook's intention to call at and establish the position of as many of the islands reported by earlier navigators as he could. The islands Tasman had named Middleberg and Amsterdam, now known as Eua and Tonga'tapu, both part of the Tongan Group, were Cook's next intended landfall.

In the forenoon of 23rd September, land was sighted from the masthead and, a couple of hours later, it was visible from the deck. It was one atoll of a group which Cook first named Sandwich Islands, but later renamed Hervey Islands. It seems that he wanted to keep his patron's name, the Earl of Sandwich, for a larger island in the New Hebrides. To confuse matters even more, the Hervey Islands have since been renamed Cook Islands.

They closed to within three miles of the island but, as they were short of time, they continued on their way without landing.

On 2nd October, at 1800, Eua was sighted dead ahead and they altered course to pass to its south. Quite soon, they spotted another island, much smaller than Eua. As they were so close to each other, Cook was concerned that they might be joined by a reef. It was getting dark, so they shortened sail and spent the night tacking back and fore.

When dawn broke on the following day, Cook and his crew were able to see that there was deep water between the islands and they turned to sail up the west coast of

Eua, which runs roughly north-west / south-east. Unfortunately, the sea was breaking so heavily on this shore that there was no hope of anchoring with safety.

Luckily, when they reached the northern half of the island they found that the coast tended to the north, north-east. This was enough to shelter them from the wind and the oncoming seas and making it possible to anchor. Once *Resolution* and *Adventure* were anchored safely, they were boarded by natives who were most friendly. When Cook, Furneaux and others went ashore to pay their respects to the Chief, they had great difficulty in landing, so dense was the throng of people. When they did get ashore, they found that the land was cultivated and cared for to a high degree.

Sadly there was no time to linger and, the next day, they were off on the short run around the south coast of Tonga'tapu, the largest island of the Tongan group, to an anchorage off its north-west corner. Wherever they went, they were given a warm welcome and the general self-assured friendliness of the populace was a delight. As they moved around the island, they were struck by the good order visible everywhere and found that the land was as well cared for, as it had been on Eua.

Sunset over Tonga, photographed from the deck of Didycoy.

Captains Cook and Furneaux, together with Mr Forster, made a last visit ashore in the early morning of 7th October to pay their respects to the person Cook believed to be the king of Tonga and to some other chiefs and dignitaries. (When we called there ourselves, the king was unmistakable, he weighed 426 pounds!)

Cook was so impressed by the warm welcome they received, that he named the Tongan group – The Friendly Isles – a name they still bear today and, if our own experience is anything to go by, they still deserve this title.

Diplomatic or social meetings with the various chiefs and other influential people may have taken valuable time, but the real aims of the voyage were never forgotten. The position of both islands was established, survey work carried out and observations on many aspects of the islands and the life of its inhabitants noted. The botanists were, as always, collecting specimens and recording the rich flora and fauna of the islands.

The acquisition of food was as crucial as ever. The quantity of victuals they picked up in Tonga'tapu is staggering. One wonders where they put it all! How do you stow 150, presumably live, pigs and 300 chickens? In tropical temperatures they surely could not be slaughtered and preserved, and what about the food for them? Forget about the yams, plantains, bananas, coconuts and the sweet potatoes they also purchased!

When Cook left Tonga on 8th of October, he set a southerly course but, whenever the wind allowed, he would turn to the west. He intended to make his way to New Zealand, contact some Maoris, in the North Island, and give them two pairs of boars and sows, cocks and hens, in the hope that they would raise them and allow them to breed, before slaughtering them for food. They would also give the islanders a variety of seeds to cultivate, thereby ensuring future food supplies.

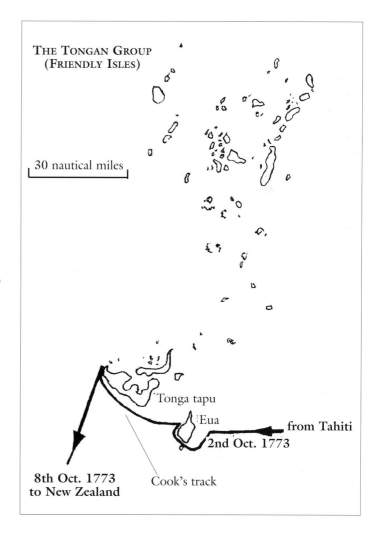

TAHITI TO QUEEN CHARLOTTE'S SOUND   149

The passage from Tonga to New Zealand was uneventful, the winds varied in strength from strong gales to gentle breezes, but always allowing them to make progress in the right direction. But the night before they sighted New Zealand, the weather turned really foul. To avoid losing contact with *Adventure* it became necessary to fire a gun at intervals and to display fires throughout the night. Their reward came, when on 21st October 1773, at about 0500, the north-eastern corner of North Island appeared over the horizon. By noon, both vessels were about twenty-five to thirty miles to the west of Table Cape, which is on the east coast of North Island.

The two vessels carried on south until about 0930 on the following day. They had passed Cape Kidnappers and were sailing across the bay beyond, when they spotted three canoes, heading for them. To enable the canoes to catch up, *Resolution* and *Adventure* hove to. The second canoe carried a Maori, whose clothes and bearing indicated that he was a chief. He came aboard without hesitation and was taken below to Cook's cabin, where he was given some nine-inch-long iron spikes, which delighted him, some cloth and a mirror. He was also given pairs of pigs and chickens and quantities of seeds. The Tahitian who accompanied Cook, explained to the chief that the animals were not for immediate consumption, but should be allowed to multiply. Similarly, the seeds and roots should be sown and cared for to produce edible crops and seeds for the following season.

The next day Cook was pressing on south in strong winds, when *Resolution*'s fore top gallant mast was carried away. The wind had piped up to a strong gale and the night was spent sailing on and off, under the fore and the main courses. The 'courses' are the lowest square sails on the two masts. In the thick fog the two ships lost contact with each other and no amount of firing of guns or displaying of fires could reunite them. The two vessels were of a similar design but because *Adventure* was smaller than *Resolution*, it was inevitable that she should be a little slower than Cook's ship and this, allied to the foul weather, would have made it difficult to stay in touch. The separation was to have no serious consequences and they headed for their rendezvous in Queen Charlotte's Sound.

By noon, on Sunday 24th, the wind dropped to a flat calm, giving the men aboard *Resolution* the chance to replace the broken top gallant mast and get the yards crossed and the sails bent in place. Almost as soon as they had completed their task, the calm gave way to yet another gale. Cook pressed on south all night, under much reduced sail and, early next forenoon, *Adventure* hove in sight. The vessels were reunited, but in very difficult conditions. The storm stayed with them throughout Monday and until midnight on Tuesday. It must have been bad because

Cook wanted to get within the shelter of the straits that separate the North and the South Islands. He had been fighting his way to windward towards the straits for some days but, eventually, he had to give up. They lowered the remaining sail and ran before the wind under bare poles, which took Cook directly away from his objective. It was not long before the two ships lost sight of each other again. This was not surprising. In a storm of this magnitude, the seas would have been so high that the hull of a vessel, when it was in the troughs, would be lost to view. The wave tops are snatched off by the wind, and driven as great sheets of dense spray and spindrift, considerably reducing visibility. The one shred of comfort for both ships' companies was that this storm blew from the north-west, so there was no danger of being driven on to a lee shore.

For a while after midnight on Tuesday, the wind became variable, both in strength and direction, as it often does after a hard blow but, by 2200 it came in from the north and *Resolution* was able to stand in towards the coast of South Island. At 0300 the wind was still blowing from the north, but they were getting close to Cape Campbell, which is on the north-east corner of South Island. Cook had *Resolution* tacked to take her away from the land and to make some progress towards Cape Palliser, with a view to getting into the straits, hopefully, on the next tack. The masthead lookout reported *Adventure* to the south, making her way towards *Resolution*. The wind strength continued to increase and although *Adventure* had been a mere three miles away from *Resolution*, they lost sight of each other again.

The wind played ducks and drakes with *Resolution* until 3rd November, when Cook, tired of beating back and fore in foul weather, put into what is now Wellington Harbour. It was about as snug an anchorage as one could wish for, protected from the wind and sea from all directions and with a sandy bottom that was ideal for anchoring. Shortly afterwards a light wind came in from the north-east and they were on their way again. No sooner had they cleared the entrance of the harbour than the wind changed direction to the south and piped up to a fresh gale that increased in strength throughout the night. *Resolution* fought her way into Queen Charlotte's Sound and anchored in Ship's Cove, but not before most of her sails had been damaged, some quite badly.

The next morning, repair work on the vessel started with a vengeance. There was much to be done before they would be in a fit state to face the next leg of the voyage that would take them back to the Antarctic.

When Cook was ready to leave, there was still no sign of Captain Furneaux's ship. Cook wrote down his intended course, had the letter placed in a bottle and buried it at the foot of a tree, pointing out the location of the message by carving: LOOK UNDERNEATH on the trunk of the tree.

The two vessels had missed each other by four days. When the storm began, Furneaux had been three miles behind *Resolution*, but rather than fighting the storm, he let the wind drive *Adventure* north and east. When the storm finally abated *Adventure* was off New Zealand's North Island. Furneaux eventually managed to fight his way back to the coast and was forced to seek shelter and to renew his food supplies. In his journal, Furneaux, hints at the terrible conditions they had faced. He describes how the decks on his vessel had become leaky, 'the people's bed and bedding wet' and how they 'began to despair of ever getting into Charlotte's Sound or joining *Resolution*.' It would take them almost a month to enter the Sound, by which time Cook was gone, on his way to the Antarctic Circle.

In his buried message, Cook told Furneaux that he intended making a broad sweep through the Pacific in search of Antarctica's coast, but also to prove once and for all that there was no such thing as the Great Southern Continent. From Queen Charlotte's Sound he would move south into the high latitudes and travel east – as far as the tip of South America – follow the prevailing winds and sail north and west on his way back to Queen Charlotte's Sound. The message suggested that *Resolution* might reach Easter Island or Tahiti around March. As he was not sure of the exact location or timing, Cook did not fix a formal rendezvous, but hinted at a chance meeting, should Furneaux follow in his tracks.

Pelican

15. New Zealand to Tahiti

November 1773 to April 1774

On Thursday, 25th November 1773, *Resolution* sailed from New Zealand, alone. Captain Cook had spent the best part of two days searching the shores of Cook Strait for signs of *Adventure*, but to no avail.

By mid-December *Resolution* was well into the Antarctic latitudes and, in the days leading up to Christmas, they suffered gales, sleet and snow, thick fog and bitter temperatures. The gear froze solid. For much of the time the ship was coated with ice, dangerous icicles hung from everything, the sails were like plates of steel and the halyards were frozen so that they resembled brittle glass rods, rather than flexible ropes. Lowering and reefing the sails was a major task. The sheaves in the blocks, through which the halyards ran, were also frozen and refused to rotate. When the wind blew with gale force, as it frequently did, the chill factor must have made the lives of the men who worked the deck an absolute misery. They were back to breaking the ice and getting rid of it over the side. And it was only mid-December, which, in the southern hemisphere, is mid-summer!

Progress was painfully slow, daily runs of twenty or thirty miles were commonplace. Indeed, on one occasion, nine miles was all they could make in twenty-four hours. With visibility ranging from poor to nil, *Resolution* had to be kept moving to give the vessel steerage way, but as slowly as possible. The proximity of icebergs, hidden in the gloom, called for great caution.

On Christmas day the weather improved, the gale abated and the sky and fog cleared. Suddenly, they could see many of the icebergs with which they had been playing blind man's buff for several days. By noon the pack ice had surrounded *Resolution* and they counted between ninety and a hundred icebergs in their immediate vicinity, 'these devilish Ice Isles,' some over two hundred feet tall. All the while, Cook scrutinised the rise and fall of the swell – there could be no Great Southern Continent in the offing.

Boxing Day was free of fog and, of course, daylight lasted all day and all night, which helped them work their way

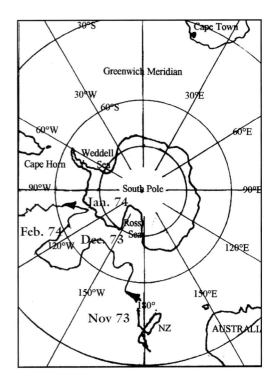

around the icy obstacles, but there was little strength in the wind and they covered only a miserly twenty miles. At 0400 a light breeze sprang up from west-south-west, which allowed them to sail to the north, in the hope of escaping from the ice, which threatened to close in on them.

By 28th December the wind still favoured a northerly course and they were soon leaving the worst of the ice behind. That day they made ninety-four miles and the next day the wind had piped up to gale force, enabling them to make a satisfying 116 miles to the north. They were still far enough south for the temperature to be in the mid-thirties Fahrenheit and the gale brought snow with it but, at least, the icebergs were becoming fewer and further between. From this position it would have been easy for Cook to sail to the east, round Cape Horn, and make his way home, but he was not yet ready to return to England. It is difficult to believe that anyone would have criticised him, had he decided to turn for home. By then, he must have had more than enough of the hardships of the Antarctic and, I suspect, his crew would have agreed with that, especially since all supplies of fresh food had long run out and they were back on dry rations. Even these were growing scarce – the men were now on two-thirds rations. They found themselves in line with New Zealand, though far out to the East, but the Cape was comfortably close. Cook, however, had other ideas.

For two weeks *Resolution*'s course was set north, usually with some east or west in it, but on 14th January he ordered the course to be changed to south-south-west and they were on their way back to the Antarctic, much to the ship's company's disappointment. John Forster expressed his bitter annoyance. Had they sailed north, they would have found rest and fresh provisions, sailing south only meant more cold and danger. As one author wrote of Cook: 'His compulsion to explore overcame the discomforts of cold and danger, and if he could stand it, so would they.' Sixteen days later, they were in the latitude of 71º south, when their way was blocked by an immense field of ice that had numerous icebergs embedded in it. There was no choice but to turn to the north again. It was none too soon, for the moment they changed tack, thick fog descended, but since they were retracing their steps, they had a reasonable degree of assurance that there were no icebergs in their path.

By the first week in February, Cook had come to a decision as to what the next stage of his voyage should be. He had considered rounding the Horn and exploring the southern Atlantic, but there was no more than a few weeks of the southern summer left and this would leave him insufficient time to carry out his plans.

To the west lay a stretch of unexplored Pacific Ocean. Unfortunately, the winds Cook had experienced in recent months had not allowed him to sail in that direction. Yet again, the persistent heavy swell from the west convinced Cook that a large landmass, such as *Terra Australis* was reputed to be, could not exist in that direction. This was evidence enough for a seaman, but it would not satisfy those who preached the gospel of the Great Southern Continent from their rocking chairs. It is difficult to think of a task requiring more self-discipline, than sailing thousands of miles in rugged conditions, in order to prove what he already knew to be true.

On the positive side, Cook knew that, up till then, the position fixing and exploration of many of the earlier discoveries in the South Pacific had been rather sketchy and this was something he was anxious to put right. One of the anomalies he wished to clear up was the position of Easter Island, as its location varied, depending on whose account you read or whose chart you consulted. Another area he wished to find and document was what he called 'Quirós's Land', the islands found in 1606 by Pedro Fernández de Quirós, and named by him: *Australia del Espírito Santo*.

Cook's plan was to be at the Horn by November of the following year. This would give him the whole of the next southern summer to explore the southern Atlantic and complete his circumnavigation.

He put his thoughts to his officers and they were quite happy with the plan, even though most of them had hoped to be heading for the Horn and home. When the men learned that they were to head back towards tropical climes, they too approved! As if to reinforce their approval, the next day produced gale force winds that brought snow and sleet in their train. Luckily the wind blew from the west, which allowed them to run to the north and sunshine.

It was none too soon. Several members of the ship's company were showing early signs of scurvy, but they would all recover when fresh food became available a few weeks later. In fact, the only person to be seriously ill was Cook himself.

Towards the end of February he suffered from severe abdominal pains and from an 'obstructed bowel,' presumably very severe constipation. He was confined to his bed for several days. In his diary, Forster says that Cook was unable to take food for over a week and that when he tried to eat, he vomited so violently, that they feared for his life.

One suggestion is that Cook had picked up a round-worm infestation. Another possible diagnosis was that he suffered from an acute infection of the gall bladder, with paralysis of the bowel. It is also suggested that, as the illness recurred later on, albeit less severely, he may have had gall stones. Clearly, trying to diagnose a medical problem 200 years after the event, and with so little evidence to go on, is pure speculation. Whatever the cause, he must have been gravely ill, so much so that the members of the ship's company wondered if he would recover. There was no fresh meat on board and Forster's much-loved dog was sacrificed. The cook made a broth of the meat and bones and it was fed to the Captain. The proteins must have helped, as he slowly recovered, but it was nearly a month before he regained anything like his former health.

For a while now, they had been close to the various longitudes that previous seamen had allocated to Easter Island and Cook wanted to determine its correct position. On 9th March 1774, *Resolution* reached the latitude of Easter Island, within three or four degrees east of one of its reported longitudes, and Cook altered course to the west. At night, some sails were furled and the vessel jogged along through the hours of darkness under easy canvas. Cook had no wish to find the island by the simple expedient of running into it in the dark. With the coming of dawn, more sails were raised and a lookout maintained at the masthead.

On the 11th they were so close to Easter Island's likely position that *Resolution* was hove to for the night. Shortly after daybreak, land was sighted from the masthead, bearing due west; by noon it could be seen from the deck. When the giant statues came into view, there could be no doubt that they had reached Easter Island!

The contacts they made with the local population proved to be friendly, but not very productive from the point of view of obtaining food and water. What water there was on offer, was brackish, and the island was clearly not as fertile as others they had visited. Add to this the poor anchorage and there was no great incentive to stay longer than five days.

Although their stay was short, it helped restore morale, which had been at a low ebb after the tough time experienced in the Antarctic. The men who were beginning to go down with scurvy, soon recovered, when fresh food was added to their diet.

THE MARQUESAS

From Easter Island *Resolution* sailed a north-westerly course. It was Cook's intention to revisit Tahiti but, before that, he wanted to check on the position of the Marquesas Islands. Álvaro de Mendaña had discovered this small group of mountainous islands in 1595, but their reported positions varied considerably.

After an uneventful three weeks of easy sailing, land was seen bearing a little south of west and the course was altered towards it. They had found a group of five of the Marquesas Islands. The largest was Hiva

A small waterfall, hidden among the lush vegetation on Hiva Oha

158 JAMES COOK

Oha – first named Dominica by Álvaro de Mendaña – the others were Fatu Huku (Hood Island), Motane (San Pedro), Fatu Hiva (Magdalena) and Tahu Ata (Santa Christina).

Their course took them between Hiva Oha and Tahu Ata. The south coast of Hiva Oha is open to the trade winds and, as a lee shore is no place to anchor a sailing vessel, they sailed on. *Resolution* rounded the northern end of Tahu Ata and turned south coming into the shelter of the island. Quite soon, they found Vaitahu Bay, where Mendaña had anchored, nearly two hundred years before.

I was wryly amused to read that when Cook attempted to enter Vaitahu Bay, he came close to losing *Resolution*, as she was struck by the vicious winds that roar down from the mountain top and vent their spleen on the hapless mariner below. She was driven to within yards of the rocks at the entrance to the bay before they were able to turn her around onto the other tack and sail away from the danger, making a second and more successful approach later on. A little more than two hundred years later, Betty and I discovered that the winds were just as violent and we had to make three attempts to enter that same bay… It seems that some things never change!

Tahu Ata had little food, water and wood to offer. So, after a few days, having established the latitude and longitude of these islands, *Resolution* moved on.

Cook's next task on the way to Tahiti was to fix the position of as many of the atolls of the Tuamotu Archipelago as he could. A straight line from the Marquesas to Tahiti runs through the Archipelago. Bougainville had sailed through this area some years before, but his position fixing was, to say the least, casual. Cook, with his scientific approach to exploration, would have regarded Bougainville's efforts as those of a dilettante. It must, however, be remembered that his knowledge of the man's writing came through translations which have been shown to be suspect. Bougainville was known to be a *bon-viveur* and a bit of a womaniser, who was probably more interested in the charms of Venus, than in her 'Transit'. These traits would not have raised him in the estimation of a rather prim Captain Cook.

On Monday, 18th April, they came across their first Tuamotu atoll. The local people were unwelcoming and there was little food to be obtained… They moved on. Incidentally, Cook does not seem to be familiar with the term 'atoll'. He refers to a

Opposite: *Portrait of the Chief of Santa Christina – one of the Marquesas islands.*

lagoon, enclosed by the ring of reefs and tiny islands, as an 'inland sea', 'the enclosed lake' or a 'salt water lake'.

The lack of swell indicated that there were other atolls and islands in the vicinity and, each day, at sunset, sail was reduced to the minimum required to give steerage way to the vessel, more sail being set as soon as it got light.

Thursday, 21st April. With the sunrise a great swell rolled in from the south, a sure indication that there was no more land in that direction. By 1000 the high land of Tahiti was in sight to the north of west. And by 2000 they were anchored in seven-and-a-half fathoms in Matavai Bay.

The author's sketch of two of the towering figures on Easter Island. Their impact on the visitor of today is still as powerful as it must have been in Captain Cook's time.

16. Tahiti to New Zealand

April 1774 to November 1774

When Cook and his men made their way ashore, they found that in the eight months that had elapsed since their last visit, conditions for the Tahitians had changed for the better. Everywhere, there were signs of increased prosperity, the houses were bigger and more comfortable. Every house had two or more large pigs tethered in the garden. Cook noted these things with satisfaction and also wrote in his log words which, to us who know of his violent death in Hawaii, five years later, sound curiously prophetic: 'Three things made [the islanders] our fast friends. Their own good natured and benevolent disposition, gentle treatment on our part, and the dread of our fire arms; by our ceasing to observe the second the first would have wore off course, and the too frequent use of the latter would have excited a spirit of revenge and perhaps have taught them that fire arms were not such terrible things as they had imagined, they are very sencible [*sic*] of the superiority they have over us in numbers and no one knows what an enraged multitude might do.' This statement, encapsulates both Cook's affection for the people of Tahiti but also his pragmatic attitude towards the thievery which, at times, strained the relationship between the islanders and the Europeans.

Cook had intended to stay for only a short time in Tahiti and then to move on in search of a location that could supply the considerable quantities of fresh food they required. They were delighted to find that Tahiti was well able to satisfy their needs. Quite soon, the usual tents and workshops were erected ashore and repairs to the vessel were put in hand.

A few days after their arrival, Cook and others went ashore, at the invitation of King Otoo. When the party reached the landing place, which was some distance from the anchorage, they were surprised to see the reception that awaited them. Several large war canoes were manoeuvring in the water, but at least 160 more were drawn up along the beach. They ranged from sixty to ninety feet in length and all of them were equipped with an impressive quantity of stone missiles, clubs and spears.

In addition, there were about 170 smaller craft, mostly catamarans, with a hut built on the deck, bridging the two hulls. These smaller craft were rigged with a mast and sails and, as many of them were laden with food, the European visitors assumed that these boats were in the nature of a supply fleet. On the shore, there was a vast throng of men, most of them armed.

It must have been a daunting sight for the visitors, but as they were there by invitation, they landed and were received with great courtesy. One of Cook's hands was grasped by a dignitary, called Towha, and the other by a man by the name of Ti'i. Unbeknown to Cook, Towha was in fact the Admiral of the assembled fleet and he was trying to pull Cook towards the canoes so that he could inspect the fleet. The situation was not at all clear to Cook. He knew that Ti'i was an uncle of the King and the Captain was anxious to honour the King's invitation, in spite of the apparent friendliness, he was also concerned at the possibility of being kidnapped by Towha. At Cook's insistence, they went off to the meeting place, as arranged, but Ti'i disappeared, presumably to find his royal nephew. Towha renewed his efforts and finally got Cook to go with him to the fleet of war canoes. Cook still felt it unwise to go aboard, as Towha wanted, and made the excuse that he did not want to get his feet wet. At this, Towha gave up and disappeared.

Ti'i reappeared and told Cook that the King was not available right now and that he, Cook, should return to his ship. The latter did, somewhat puzzled by the whole business, but the whole thing soon turned farcical. Once back on board, Cook was told that the King had fled because he feared his visitor's wrath, when he learned that some of his clothing had been stolen from the washing line, erected by the ship's laundryman.

Cook then discovered that Towha was in fact the Admiral of the war fleet which was by then on its way to chastise a dissident chief in a distant part of the island. His intention had been to do Cook the honour of reviewing the fleet, before they left. Clearly, the latter would have been delighted to learn more about the canoes and the way they were used in battle, but a succession of misunderstandings had put paid to that.

In rapid succession, Cook had failed to meet the King, offended the Admiral of the Fleet, missed the chance to learn more about the local fleet and the way it was used in battle and even lost some of his clothing... It really wasn't his day! Later on, he was able to make amends by entertaining both the King and the Admiral on board *Resolution*. The Admiral was shown all over the ship and soaked up all he saw, in much the same way as Cook would have done, had he not missed his chance to do so.

By the middle of May *Resolution* was heading westwards again. Short stops were to be made at Huahine and Raiatea in order to renew old friendships and pick up whatever fresh food was available.

Opposite: Tahiti Revisited *by William Hodges.*

At Huahine they found that Ori was still the Headman and that he was as friendly as ever. His subjects, however, were not entirely welcoming, even to the extent of attacking and robbing the occasional small group of Englishmen, if they were too far to call on others to come to their aid.

Forster's servant was one of those attacked. Forster, who always worked on a short fuse anyway, was so incensed that he wanted to shoot one of the culprits to teach the remainder a lesson. As ever, Cook was adamant that no such steps should be taken. He argued that these people probably had a different set of values from those of the ship's company and that he was prepared to do his best to avoid bloodshed. He was not above using a show of strength, but he would not willingly injure or kill the people of the islands he visited. Some of his officers and the scientists did not always see eye to eye with his compassionate approach, but he was the Captain and his word was law.

Raiatea was their next stop, where they stayed for a short time, leaving on 4th of June and heading out to the west, in the direction of the islands discovered by Quirós in the sixteenth century. As mentioned before, Cook was anxious to confirm their position. The Tongan Group also lay in his path and he had decided to visit Nomuku, that had been named Rotterdam Island by Tasman. It was the island Cook had missed on his earlier visit to Tonga.

For a while after leaving Raiatea, Cook had to be cautious as there were a number of islands and atolls in his path whose positions were uncertain. The day after they had left Raiatea, about an hour before sunset, they had Bora Bora to the north-east and Mauru to the north-west. It was decided to heave to for the night, as these small islands tended to come in clusters. At first light sails were set and they were once more on their way to the west.

The same cautious procedure was adopted for the night of the 6th and, at 1100 the next forenoon, they sighted Howe Island, a low-lying atoll, dead ahead of them. There was one more known hazard before they were in the clear: Mopelia Atoll, but they must have passed without seeing it.

Cook makes frequent reference to the large swells coming from the south. They were an indication that no large landmass could lie in that direction, or the swells would not have built up. On the other hand, there were many sightings of tropic birds and frigate birds, also known as man-of-war birds, a reliable indication of land in the offing.

The frigate birds are normally only found no further than some fifty miles from land. They are powerful birds, with a wingspan of six feet or more, and, whilst their diet is fish and still more fish, they do not normally catch it themselves. They tend to accompany other sea birds and, as soon as one of them has caught a fish and is heading for the shore to feed its young, it is harassed until it drops its catch, which is promptly gathered up by the nearest frigate bird.

I have been on the receiving end of their harassment techniques and it is pretty fearsome. I'd caught a large dorado and had taken it onto the fore deck to scale and gut, when about six frigate birds decided that they wanted my catch. I clutched the

Opposite. *We took this picture of a* frigate bird. *A German couple had rescued the bird, who had a damaged wing and was unable to feed itself. They had to become proficient fishermen in order to keep their patient in fish. The wing mended, but the bird refused to leave their ship and the luxury of regularly served meals.*

fish to my bare breast and beat a hasty retreat to the safety of the cabin. To be the centre of interest of a mass of thrashing wings, huge hooked beaks and outsize talons is an experience that is not easily forgotten!

Back aboard *Resolution*, on the 16th of June, large numbers of frigate birds appeared and, soon after, a low-lying atoll was sighted. Cook gave it the name of Palmerston Island in honour of Lord Palmerston, one of the Lords of the Admiralty. They could see no opening to the enclosed lagoon and no sign of habitation, so they sailed the length of the shoreline, noted its position and resumed their course.

Five days later land was seen to the west but, as it was only an hour or two before sunset, they reduced sail and waited for daylight. As *Resolution* approached the coast the next morning, people were seen near the shore but they quickly disappeared into the nearby forest. Two boats were lowered to the water and Cook, with a number of officers, some of the scientists and an armed party of marines, went ashore hoping to meet some of the islanders. The locals showed considerable hostility to the landing party and it was obvious that no good would come of further attempts to make contact, so the shore party returned to *Resolution*.

Later research has suggested that the menacing reception accorded to Cook and his companions was, in fact, a ceremonial challenge that would have preceded a welcome. That may be so, the Maoris do have a similar tradition. Cook, however, did not linger to find out. He named what is now called Niue, Savage Island, and went on his way.

By the evening of Saturday the 25th, they judged themselves to be close to the position of Rotterdam Island or, to give it its Polynesian name – Nomuku – one of the Tongan Group. Sail was reduced to the topsails for the night.

With daylight, a number of islands came into view, extending obliquely across their course. As *Resolution* drew closer, they saw that a reef associated with the islands barred their way and they turned to the south to find a route round the obstruction. At last, on the following day a suitable passage between the nearest island and the reef was spotted and they turned into it. The leadsman reported forty to forty-five fathoms of water and the tallow on the base of the lead showed that the bottom was sandy. This was good news, for it meant that they could now anchor in an emergency. Still, it was a lee shore and no place to anchor from choice. An hour or two before sunset on 27th of June, *Resolution* had worked her way round to the north side of Nomuku, and anchored. They had twenty fathoms of water, a sandy bottom, shelter from the south-east trade winds and were within reasonable distance of the shore with the promise of fresh food. What more could a sailor ask for? Girls in hula hula skirts? Well, they were there too!

Early next morning, Cook and the scientists went ashore, along with others. Some of the working parties were busy collecting wood and water, while others were bartering with the locals for fresh food, which was in plentiful supply. The cove they used to reach the shore dried out as the tide ebbed away and all this activity had to end at midday. Everyone, except the Surgeon, noticed what was happening and returned in time to *Resolution*.

After dinner, some of the officers went ashore and found the Surgeon, unharmed but looking somewhat sheepish. Earlier that day, when he realised he had missed the boat back to *Resolution*, he had asked the occupants of a canoe to take him back to the ship, but no sooner had he stepped into it, that the islanders took his musket and made off with it.

As soon as Cook learned of what had happened, he went ashore with an armed party in two boats. He was concerned that some of his officers would act precipitously, even to the extent of causing loss of life. When he found the shore party, there were quite a number of Polynesians on the edges of the group. Once he knew how the Surgeon had lost the musket, Cook felt very strongly that the Surgeon was to blame for its loss. No attempt had been made to recover it and Cook decided to take no action.

Next morning, a watering party made their way ashore, only to find themselves surrounded by a number of local people, who were less than friendly, so much so that the crew were in some doubt about collecting the water for which they had come ashore. However, they knew that Cook would soon be there with an armed party of marines, so they pushed their way to the river, filled the casks with fresh water, and then, with some difficulty, got them back to the boat. While they were busy with these tasks, the natives used the opportunity to steal another musket, some of the cooper's tools, and personal items from a number of sailors.

When Cook reached the scene, he was received with apparent courtesy by the Polynesians. On being told of the thefts, he made his displeasure known and one musket was returned. Cook was followed closely by a boatload of armed marines and, when they stepped ashore, alarm spread among the local people and many of them fled. The marines seized two large catamarans in reprisal and, one man who resisted, was peppered with small shot and limped away. Not long after the catamarans were seized, the second musket was returned. The catamarans were then released and all of *Resolution*'s people returned to the ship. Later, the cooper's adze was also returned.

Cook's conciliatory action over the loss of the Surgeon's musket the previous day had obviously been seen as a sign of weakness and only encouraged the locals to steal other items. In his log, Cook blamed himself for the incident and says something to the effect, that he should have taken firmer action, when the first musket was stolen. The man who had been hit by small shot, was seen by the Surgeon who dressed his wounds, that were superficial and healed quickly.

Whilst ashore, Cook learned that there were some twenty other islands, scattered from the north-east to the north-west of Nomuku. Some of them could be seen from Nomuku itself. One appeared to be an active volcano but, when they drew close to it a few days later, they found that the smoke they had seen was probably the natives burning off scrub, before planting the ground with new crops.

After leaving Nomuku, *Resolution* ghosted along in a gentle breeze for several days, often moving so slowly that Tongans in canoes caught up with them and bartered fresh food in exchange for the usual gewgaws.

Sailing so slowly among the numerous boats that came out to meet *Resolution* from the islands they were passing, gave the Europeans the chance to observe the way these craft were sailed. A European sailing vessel, when changing tack, would either steer its bow or stern across the wind and adjust the sails so that they would fill from the other side of the vessel. The Polynesians achieved the same result by bringing the boat's sail around the mast, so that what was the bow became the stern and vice versa.

It is in this area that Captain Cook first mentions having seen people with leprosy, but he could well have been mistaken in his diagnosis. There is a tropical disease called yaws, which produces rodent ulcers that eat into the flesh and, if they manifest themselves on the face, they can destroy the nose and other areas of soft tissue. Cook could be forgiven for mistaking this condition for leprosy, because the symptoms are sufficiently alike to fool a layman. Nowadays, a single injection of long-acting penicillin is usually enough to arrest the progress of the disease, but two centuries ago, there was no known cure.

On Sunday, 3rd of July, land was seen from the masthead and Cook had *Resolution* steered towards it. As they got closer, they found it to be a very small island with a fringing reef. Several turtles were seen near the reef, which prompted Cook to name the island, Turtle Island. He noted its position and moved on. Its reduced size and fringing reef persuaded them that there was nothing to be gained by stopping.

The next twelve days were spent sailing cautiously to the north-west. Each night, sail was reduced and then restored at daybreak. Most days, until 15th July, when they reached the latitude of what Cook called Quirós's Land, there were signs of land: tropic birds, frigate birds and seaweed. Two days later, in the afternoon, they saw land to the south-west and, by 1930 that evening, they had reduced sail and tacked to move away from it. At 0200 they came about again and headed in towards the island.

It must have been quite a rough night with gale-force winds from the south-east. Several sails were badly split and others were damaged beyond repair.

When daylight came on 18th July, they were close to the northern end of what would be called the New Hebrides. This is a string of islands that lie in a line, from a little west of north to a little east of south, for about six hundred miles. Quirós had named them *Australia del Espíritu Santo* and Bougainville, some time later, called them the Great Cyclades. They are now known as Vanuatu.

Their landfall was the long narrow island of Maewo, at the northern end of the chain of islands. As soon as it was light enough to enable them to see the hazards that lay around them, Cook had the double-reefed topsails set and *Resolution* battled her way northwards against the strong winds, until they were able to round the top end of the island. Once they were in its shelter, the seas were no longer a problem, but the winds were as strong as ever. They spent the next day, tacking back and fore between the islands of Omba and Maewo, making their way to the south.

With such strong winds on the nose it was a slow task, tacking back and fore repeatedly, gaining a little distance in the right direction with each tack. A vessel like *Resolution* would probably not sail closer to the wind than seventy degrees and to make six-and-a-half miles to windward, it would be necessary to sail twenty and maybe more, if the weather was really foul.

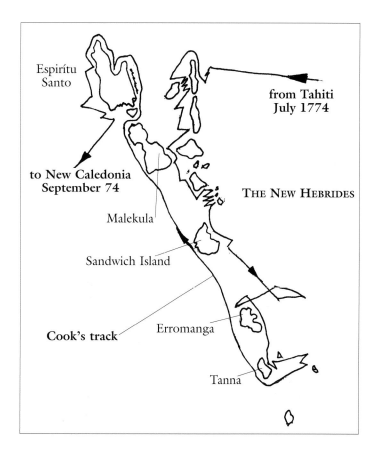

As mentioned before, Cook was using an English translation of Bougainville's account of his voyages, which also contained the charts he had made of his discoveries. When Cook was able to establish the position of these islands, he found their charted position to be twenty miles too far to the north. As this was the same error as in Bougainville's charted position of Tahiti, Cook concluded that the mistake was Bougainville's and suggested that all of his Pacific positions could well contain the same degree of inaccuracy. He may have been right but, to be fair to Bougainville, the error might have been in a faulty translation. If the error was Bougainville's, it was a surprising one to make, because determining latitude was a fairly simple matter, unlike longitude.

TAHITI TO NEW ZEALAND

Friday, 22nd July. Cook found a secure anchorage on the south-east corner of Malekula. They were received in a friendly, but not unduly welcoming manner. There was a very limited amount of food to be obtained and, apart from a small amount of firewood, it yielded little of use to the crew of *Resolution* and they were away at 0700 the next morning.

In all, they anchored only three times off the islands of the New Hebrides and, on each occasion, their stay was a short one. Cook had no time to linger at this stage of his voyage, as he was anxious to chart as much of this part of the South Pacific as possible, before returning to the Antarctic and making use of the southern summer. His immediate interest was to determine the extent of the New Hebrides group of islands and he directed his course to the south.

From Malekula, their course took them as far south as Erromanga, on which there was an active volcano. On the eastern side of the island, there was a bay which offered shelter from the seas generated by the southerly winds and *Resolution* was anchored there. Cook's first landing on Erromanga was made in the face of a vast crowd of armed natives. He carried a green branch, which he used as a symbol of peaceful intentions, throughout the South Pacific. It is not sure what the locals made of this gesture! The reception was cool, but courteous. Cook was unaware that the islanders saw his white skin as a sign that he and his companions were ghosts. He indicated that he would like to get some food, meaning food for the ship's company. This was interpreted as a sort of ceremonial offering and he was given a little fruit and water, which he accepted. He then attempted to return to the boat, intending to get back to the safety of *Resolution*. This seemed to be the signal for the crowd to attack. Arrows were shot and spears and stones thrown at the men in the boat. Muskets were fired at the crowd and eventually the boat was able to head out to *Resolution*. The damage to life and limb was limited to the death of the local Chief and to three or four minor injuries on both sides.

It was decided to shift *Resolution* to a position closer to the landing place, to allow her guns to be brought to bear, should they be needed to protect future shore parties. As soon as the anchor was free, the wind came in from the north, a direction that turned the bay into a lee shore and it was no longer a tenable anchorage. Cook ordered the sails to be set and sailed out of the bay. For the rest of the day, and part of the night, they sailed to the south, guided by what they thought was a great fire. By 0100 they were close enough to the shore to shorten sail and make short boards, on and off, for the rest of the night. With the advent of daylight, it became apparent that the fire they had seen throughout the night was in fact an erupting volcano, throwing up large volumes of smoke and fire, laden with rocks and lava. The rumbling it made could be heard from a considerable distance. Later, Cook and the Forsters, along with others, attempted to get close to the volcano, but were defeated by the distance from the anchorage and the obvious disapproval of the local people.

This was Tanna Island and a suitable bay in which to anchor was soon found. The ship's boats were put into the water to go ahead of *Resolution* and check the depths. The boats signalled that these were sufficient to anchor and *Resolution* made her way in. Once she was in the entrance of the bay, the wind deserted her and she was obliged to anchor in four fathoms. While she waited on her anchor, the boats made further checks on the depths and, when Cook was satisfied, kedge anchors were laid and *Resolution* was warped into a suitable spot, where she anchored.

At first the locals stayed well clear of *Resolution* but, slowly, they overcame their caution and started to trade food for trinkets. Soon they became bolder, to the point where they became bold enough to steal anything they could lay their hands on – the anchors' tripping lines and buoys being prime targets.

When the anchors were laid, it was customary to attach a tripping line to the lower end of each one, with a buoy bent to the other end of the line. Should the anchor become jammed under an obstacle on the seabed, the anchor cable could be slackened off and the tripping line hauled in, pulling the anchor out upside down. It was these lines and buoys that so appealed to the thieves.

The next morning, Cook wanted to go ashore and three boatloads of armed marines and seamen were made ready. On the foreshore, some thousand natives had assembled, all armed. *Resolution* was in such a position that the beach, and the forest behind, formed easy targets for the muskets and the heavier guns she had on board. One fellow was unwise enough to bare his backside to the ship and beat on his buttocks with his hands, making a perfect target for a charge of buckshot, from a musket fired by the Lieutenant of Marines.

Resolution stayed at Tanna from 5th to 19th August. Slowly, the people became less hostile, but never to the point when the Englishmen felt they could relax in their presence. Apparently, the inhabitants of Tanna too, thought that these white men were ghosts. They kept to themselves and Cook and his men thought that the relationship was improving. Towards the end of their stay, however, a sentry who believed his life to be endangered, killed an islander with a musket shot. Cook had the sentry arrested and put in irons, with the promise of a flogging later. The officers, some of whom were present when the incident occurred, argued strongly against the man being flogged and, after some heated discussion, Cook accepted their arguments, but insisted that the man should remain in irons for some time. The Midshipman, John Elliot, taxed Cook of having lost sight of 'justice and humanity.' The object of the fracas, the sentry, was to die soon after, anyway, falling overboard when drunk.

During the night of 20th August, a favourable wind blew in from the south, south-east. Resolution raised her anchors and sailed to the east to get a good offing, before attempting to round the southern end of Tanna. The two remaining islands of the southern end of the chain were well in view, so Cook could turn to the north, as soon as he was clear of Tanna.

When they left Tanna, *Resolution* sailed up the western side of the island chain until she reached the northern end of Malekula. From here a course was laid that would take them through the Bougainville Strait, which lies between Espíritu Santo and Malekula.

On 1st September, Cook was sailing away from The New Hebrides, having completed a running survey of the archipelago. In addition, notes of the physical geography of the islands, the precise position of each of them, details of tidal behaviour, depths of water and the magnetic variation in various locations, were all recorded. The frequent calms that *Resolution* endured, as she circumnavigated the group, enabled Cook to gauge the strength and direction of the currents at various places around the islands. All this, and more, achieved in a mere forty-six days. When one remembers that Cook recorded the same data, wherever he met land that was new to him, one begins to realise the extent of the massive contribution he made to the store of knowledge of the world, in what was a comparatively short lifetime.

Land was sighted from the masthead on 4th September. It was estimated to be about twenty-five to thirty miles to the south, south-west and extended as far as they could see, in either direction, across their course.

As they drew nearer, some openings appeared in the coastline, but they could not decide whether they were inlets or open water. A headland came into view which Cook named Cape Colnett, after the man who first spotted it. By 1800 *Resolution* was about nine miles off and becalmed. It was some hours before a gentle breeze set in from the south-east and they spent the rest of the night sailing slowly off and on, waiting for daylight.

Come sunrise, they could see a fringing reef along the length of the coastline. Cook chose to turn to the north-west, cruising about five or six miles along the shore and it was not long before they found a passage through the reef. Two armed boats were lowered to the water to check the depths in the passage and in the lagoon beyond. Whilst the boats were being prepared, ten or twelve sailing canoes came out towards *Resolution*, but retreated when they saw the boats being lowered.

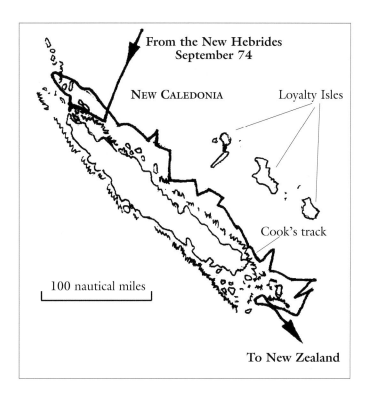

The boats reported a navigable channel and sufficient depths of water within the lagoon for *Resolution* to lie at anchor. With the two boats lying in positions that marked the channel, *Resolution* made her way in and anchored in five fathoms, on a bottom of fine sand mixed with mud. Surrounded as they were by land on one side, and the fringing reef and small islands on the other, they were in a near perfect harbour.

Cook and those who went ashore were treated with courtesy and civility and although they stayed for ten days, the good relations never faltered. Fresh water, wood and a moderate amount of food were available. A present of a hog and a sow, a dog and a bitch, was made to the headman. Dogs were normally eaten, in many of the Polynesian islands, but Cook suggested that these dogs and pigs should be allowed to breed, before they were eaten.

The scientists were particularly pleased to be able to roam so freely and made the most of the opportunity. The astronomers were able to check the chronometers and establish the longitude of the island, as there was an eclipse of the sun whilst they were at anchor.

It took Cook a little over two weeks to circumnavigate and chart New Caledonia, which is about 250 miles long by thirty-five wide, making it, after New Zealand, the second largest island in the Pacific. Having established that New Caledonia was an island, Cook demolished one more myth, proving that it was not part of the mainland of Australia, as had been claimed by some geographers.

They left New Caledonia with a strong north-westerly on their tail and made good time to Queen Charlotte's Sound, in New Zealand, arriving there on 18th October 1774. It was obvious that another ship, presumably *Adventure*, had been in the Sound, the message left for Furneaux had gone and trees near the anchorage had been cut down fairly recently, using a saw and axe. The ship was long gone and Furneaux had left no return message. For at least a week, the Maoris were conspicuous by their absence and this was in an area where they had been friendly before. Cook finally made contact with them, when he was out shooting birds for the pot. The Maoris were seen at some distance and they were reluctant to approach Cook, until they were sure it was indeed him, then they ran to him and greeted him effusively. They promised him that they would bring fish to barter the next day, which they did.

When questioned about *Adventure*, the Maoris were reluctant to talk about her. In the end Cook hit on the idea of drawing an outline of Queen Charlotte's Sound and cutting two paper shapes to represent *Resolution* and *Adventure*. The Maoris caught on at once and Cook learned that *Adventure* had been and left, but that was all. The Maoris seemed to be holding back information and, to his concern, there was no way he could get them to elaborate.

Tropic bird

17. New Zealand to England via Cape Horn

November 1774 to July 1775

Resolution was ready to sail on 10th of November and, as the southern summer was upon them, Cook was anxious to be away. The barque was moved out of Ship's Cove and anchored so as to be ready to leave as soon as the wind was favourable. To Cook's satisfaction, conditions were right on the very next morning and they were off.

Once they cleared New Zealand, the course chosen was as close to south as the wind would allow, only diverging to the east when the wind forced a change of course. By the end of November, they were well into the higher latitudes and heading towards Cape Horn. The weather they experienced varied, but was seldom good and gales were commonplace. And, of course, it was cold!

It must have been a somewhat frustrating time, as Cook had long been convinced that there was no possibility of a Great Southern Continent being found in the South Pacific, and that his time could be spent more usefully elsewhere. Yet his orders required that he completed the circumnavigation in those latitudes in which, it was claimed, *Terra Australis Incognita* might be found. The last leg of the voyage required him to sail from his present position to Cape Horn and then into the South Atlantic, as far as the longitude of the Cape of Good Hope.

On 17th December Cook altered course towards Cape Deseado, which is at the north-west corner of Desolation Island. East of this Cape is the Pacific entrance to the Strait of Magellan. Cape Deseado is also the northernmost point of the archipelago, that forms the western side of the Strait.

At midnight, land was sighted and *Resolution* turned to the south-east to run parallel with it. The weather had become variable, ranging from gales to calms and back again. They followed the coastline for three days, with Cook and his team taking copious notes, as they made their way to the south-east. Cook frequently comments in his journal on the barrenness and the desolation of the land.

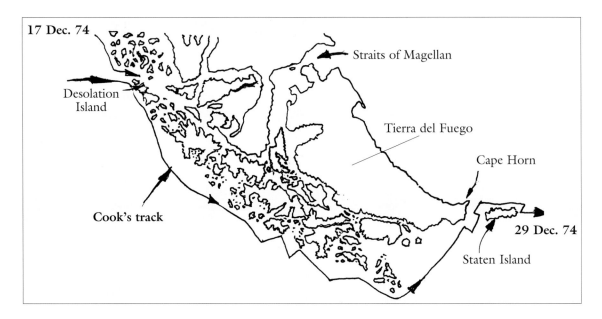

This is indeed a part of the globe that looks like the end of the world. From the ship they saw a coastline, deeply indented with watery inlets that led nowhere. There was a succession of small islands that seemed to consist of craggy pinnacles of rock, totally devoid of vegetation, or of any other form of life. Later, when they were able to get ashore, they discovered that there were some valleys containing small pockets of land, with a few trees and other meagre vegetation.

Cook likens the archipelago to the Norwegian coast and says that, in the time at his disposal, it was impossible for him to carry out a detailed survey of the coastline, due to its convoluted nature. He expresses his belief that, for all practical purposes, there was now sufficient information to help other ships and that he must leave it at that. What Cook had put together was, in fact, an invaluable set of sailing directions, that would enable other masters to sail safely in these dangerous waters.

Cook decided to enter one of the fjord-like openings in the coastline, in the hope that they could replenish their water supply and, perhaps, also find some wood, as their stocks were running very low. When they got inside the bay, the water was so deep that they could not reach the bottom, although they were using a lead-line of 170 fathoms.

Islets were scattered around and nightfall was not far off. Cook pressed on, his desire to set foot in this new land, for once, overcoming his prudence. Had he failed to come to an anchor before darkness set in, his position would have been fraught with danger. Fortunately, his luck held and a small cove came into view. The boats, that had been sent to sound the depths, came back with the welcome news that the water within the cove, was about thirty fathoms deep.

As soon as they were in, the anchor was let go, a kedge was taken out astern and they were secure for the night. The next morning two boats were sent out in the hope of finding a more secure anchorage. It was not long before they returned, with the news that there was indeed a better anchorage, a short distance off, which offered both fresh water and a small valley with a stand of trees. The new anchorage was secure from winds from all directions and the wood and water parties were soon at work. The astronomers set up their observatory and the botanists were delighted to be ashore and able to roam among pastures new. Although they had seen no sign of people in any of the islands they had sailed past, when they got ashore, they found the remains of fires and a hut. They thought it prudent to land a party of marines, who would keep watch over the gear brought ashore by the various work parties.

Towards the end of their stay, contact was made with a few of the local people, some of whom came aboard *Resolution*. In the reminiscences of some of the ship's crew, the locals were described as dirty and wretched, inadequately dressed for the climate and they were not encouraged to stay on board.

It seems to have been a happy break for Cook. He was free to wander, hunt wildfowl and generally explore the area. Christmas was celebrated, roast goose was on the menu for all hands. They stayed for three days, then, they were on the move again.

At 1930, on 29th December, *Resolution* rounded the Horn and sailed into the South Atlantic. Two days later, they found themselves off the north coast of Staten Island, in thick fog. Sail was reduced to the three topsails and *Resolution* was allowed to jog slowly along until the fog cleared. Cook wanted to get ashore on Staten Island, so they moved on towards the eastern end of the island, seeking an anchorage that would give them some shelter, but allow them to sail off, should the wind come in from the wrong quarter. By 2000 they had Observatory Island abeam and Cook decided to sail round the eastern end and anchor somewhere off the island's south shore. There is a rip tide at the eastern end of Observatory Island and *Resolution* sailed through the rough water it kicked up, with the leadsman giving them a depth of nineteen fathoms, which was more than enough depth for *Resolution*. Cook found his spot and had the vessel anchored in twenty-one fathoms, at no great distance from the shore, with Staten Island on the other side, giving them some shelter as well. The anchorage was not ideal, though, as the ground was stony.

The whole of the beach was covered with seals and birds and, once the vessel and the crew were settled, three boats were sent ashore to kill some seals and birds and gather what fish they could to supplement their diet. The blubber of the seals and sea lions proved very useful: it was rendered to make oil for the lamps.

The next morning the Master, Mr Gilbert, was sent in the cutter to see if he could find a better anchorage and, by 1000, he was back with good news. He had found a sheltered spot, with a sand and mud seabed that would offer much better holding than the present anchorage with its stony bottom. There too, the shore was home to great numbers of seals; geese, ducks and various other birds and there was plenty of wood for the galley stoves.

Unfortunately, the day was too far advanced and the wind too weak to allow them to shift *Resolution* that day. The next day the wind was fitful until late in the day. Very early on Tuesday morning, a fresh breeze came in from the north-west and *Resolution* set sail for the new anchorage. But it was not to be, the wind turned squally and then fell away to a calm. When it eventually returned, the wind was not strong enough to drive *Resolution* against the current and she lost ground to it.

Eventually, the wind did improve and, on rounding the eastern end of Staten Island, they met with another rip tide that gave them a rough ride. Once they had negotiated the tip of the island, the wind strength increased to the point where they had to double-reef the topsails. Given time, Cook was able to get ashore at Staten Island, but what he saw did not impress him. There was nothing but rocks, and the surrounding waters were so deep that anchoring would be problematic.

Cook was now in the Atlantic and on the last leg of his search that would prove or disprove the existence of a Southern Continent. He directed his course to the east, with the intention of sailing to the longitude of Cape Town, thereby completing his circumnavigation of the world in these latitudes.

On 13th January, they sighted what, at first, they thought was a huge iceberg but, as they drew closer, it became apparent that it was part of a range of snow-covered mountains, which stretched across their course. *Resolution* approached, rounded the northern end of the unknown land and sailed along its east coast for several days. As Cook sailed south-eastwards, he did what surveying he could, but the work was hampered by dense fog and gale-force winds, interspersed with calms. When they reached the southern-most point, they turned to the south-west and sailed along the south coast until the northern end of the island came into view, proving that it did not form part of a much larger continent.

The land they had found was what we now know as South Georgia. It is possible that others had been there before Captain Cook, but the evidence is inconclusive. He went ashore in three places, named the land the Isle of Georgia and claimed it for King and country.

Wednesday, 25th January 1775. *Resolution* was heading east, south-east. By Friday they were close to 60º south and in fog so thick, that they could not see a ship's length. Cook was tired of these cold and foggy latitudes and decided to seek what land might be found near Cape Circumcision. For nearly a week *Resolution* tried to make her way to the north, but the wind did not favour them and they were forced to sail a course, that was much more to the east than to the north. Repeatedly, they were enveloped in fog.

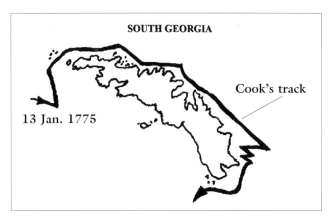

Early on the morning of 31st January, the fog cleared finally to reveal land, three or four miles ahead of them. By the time they had altered course to miss the obstruction, they were within a mile-and-a-half of the breakers. When the visibility improved, they found that there were three small, but very steep, islands in their path. They identified them as the Sandwich Islands. The water around them was far too deep to allow them to anchor and, the few small bays they could see, were all choked with ice. As soon as it was safe to do so, they altered their course northward and maintained it for 180 miles or so, passing six or more islands on the way.

Saturday, 4th February. They were once again on an easterly course, about 58º south, back into snow and ice. Not for the first time, they encountered a very strong gale that was combined with an exceedingly heavy fall of freezing snow. The gale caused the snow to pile up on the sails, making the ship top-heavy and unstable. Again and again, they found it necessary to point *Resolution* very close to the wind, so that her sails lost their shape, but shook so hard in the gale-force winds, that the snow was thrown off.

Cook finally decided that there was no point in searching further for the Great Southern Continent in these polar latitudes. As he rightly pointed out, the only land that could possibly be found in those latitudes must be ice- and snow-bound all the year round, and that could not be the *Terra Australis* he had been sent to find. Cook's aim now was to locate Bouvet's Cape Circumcision and settle whether it belonged to an island, or was part of a continent.

By 18th February, *Resolution* was close to the latitude of Bouvet's discovery. For three days Cook sailed mainly to the east, without finding land. He concluded that if Cape Circumcision existed, it could only be part of small island, but he was inclined to believe that what Bouvet had seen was, in fact, an iceberg shrouded in mist.

The longitude they were in completed Cook's circumnavigation in the higher southern latitudes and proved that the Great Southern Continent was a myth. This, with the charting and survey work he had carried out, fulfilled the orders he had been given by the Admiralty. *Resolution*'s gear was worn out and there was neither spare rope nor sailcloth to replace it. Breakdowns were a daily and sometimes an hourly occurrence and, this alone, was reason enough to think of the return journey.

The health of the ship's company had so far been good but, as they no longer had fresh food at their disposal, it could not be long before the dreaded scurvy set in. Clearly, it was time to go home. They were close to the longitude of South Africa and *Resolution* was turned to the north, heading for Cape Town.

Apart from the weather, the passage to the Cape of Good Hope was uneventful. As they drew closer to their destination, they met with a Dutch ship and then an English vessel, both of which gave *Resolution* welcome gifts of fresh provisions. Unfortunately, they also gave Cook a chilling report on the whereabouts of *Adventure*. The ship had made it to the Cape a year previously. Furneaux was reasonably well, but his ship's company was sadly diminished. This is when the reticent behaviour of the Maoris, which had puzzled Cook during his last visit, suddenly made sense. He learned that, while in New Zealand, a boat party from *Adventure* had been attacked and that the men had been killed and eaten by the locals. Cook had a great affection for the Maoris. They had welcomed him and he regarded them as friends. He knew of their spasmodic acts of cannibalism, but had never imagined it could affect the crew of one of his ships. He also knew of Furneaux's shortcomings and decided to wait to know the full facts, before making up his mind. He wrote: 'I shall make no reflections on this melancholy affair until I hear more about it. I shall only observe, in favour of these people [the Maoris], that I have found them no more wicked than other men.'

On Tuesday, 21st of March, Table Mountain appeared over the horizon and, later that day, *Resolution* anchored in the bay.

In the foreword to this book, I commented on the difficulty to decide the exact dates of events. Cook's arrival at Cape Town illustrates the point. According to his log, he arrived on Wednesday, 22nd of March 1775. Ashore, the date was in fact Tuesday, 21st March. Some time ago, Cook had crossed the International Date Line on an easterly course and should have deducted one day to keep his calendar in step with the rest of the world. Chances are that the dates in his log had been in error for 160º of longitude, nearly half way round the world. If you add to this the fact that, ashore the new day starts at midnight, but at sea it starts at noon, except that at least on one occasion, Cook uses midnight as the start of a new day – reverting later to using noon – gives an idea of the scope for confusion!

In Cape Town, a Captain Newte, of *Ceres*, an English East India Company's ship, was about to leave for England and he agreed to take a copy of Cook's journal, his charts and drawings and ensure that they reached the Admiralty.

Letters from Captain Furneaux awaited Cook at Cape Town, in which Furneaux confirmed the loss of eleven men and a boat in Queen Charlotte's Sound. There had been some sort of fracas ashore, during which the men were killed. He also outlined his route home. At the Cape, Cook met the captain of a French Indiaman. They found that they had a lot in common and soon became friends, comparing experiences, maps and notes from their travel. Captain Julien Marie Crozet told Cook of the voyage made by Marion du Fresne, with whom Crozet had sailed to New Zealand. Marion du Fresne and some of his sailors had been killed and eaten by the Maoris. Crozet had escaped, sailing to the Philippines, then home via Mauritius. This made Furneaux's grisly experience part of a pattern.

As soon as the courtesies with the Governor and other dignitaries were completed, Cook's priority was to purchase fresh food for all on board. Those officers, who could be spared, and the civilians, were housed ashore and the crew were given shore leave, a watch at a time. Meanwhile, Cook made yet another unpleasant discovery. He was handed a copy of 'his' journal, kept during the *Endeavour* voyage, which had been published without his knowledge and during his absence. More of this later.

Work was started on *Resolution*'s much needed refit and lasted for a month. Her rudder was found to be so damaged by the depredations of the teredo worm, that it had to be unshipped and rebuilt. Caulking materials were difficult to find and this caused further delay. But, at last, provisioned, watered and wooded, her leaky seams recaulked, her sails and rigging renewed or repaired, her rudder rehung and her crew refreshed, she was ready for the last leg home.

On the morning of 27th April, Cook and his officers paid their respects to the Governor and, as soon as they were back on board, *Resolution* sailed towards St Helena, in company with *Dutton*, an English East Indiaman, arriving on the island on 15th of May. There, Cook was to enjoy the warm hospitality of the Governor.

On 21st May, *Resolution* left once again in the company of *Dutton* and, three days later, the vessels went their separate ways. A further three days and *Resolution* was anchored in Cross Bay, Ascension Island. They had come to catch turtles to restock their larder and, within a few days, they had caught twenty-four, each weighing between 400 and 500 pounds. There was fresh meat on the menu once again.

Their next stop was at Fayal, in the Azores, to take on water and purchase fresh food. They stayed until 19th of July 1775. Ten days later they made their landfall at Plymouth and the next morning they were anchored at Spithead. Without delay, Cook, the Forsters, Hodges, Wales and a couple of others made their way to London. As usual, Cook rushed off to the Admiralty to report. There was great excitement at his return and he had a lengthy meeting with the Admiralty Lords. Daniel Solander, who had taken part in the *Endeavour* voyage, found himself at the Admiralty on that day. He wrote to Joseph Banks how he had hardly had a chance of a word with Cook, due to the length of the latter's interview with his Admiralty masters. As usual, Cook was anxious to get back to Mile End and Elizabeth. It must have been a passionate reunion, within weeks, Elizabeth was pregnant again. The joy of reunion would have been tempered, as Cook discovered that his son, George, had lived for only four months, dying in October 1772, while Cook was in Antarctica. Luckily, his two surviving boys were thriving: twelve-year old James was attending the Naval Academy in Portsmouth and Nathaniel would soon emulate him.

Cook had much to be proud of. Besides his immense scientific achievements, in a voyage of a little over three years, he had lost four men. One had died of tuberculosis, presumably already infected when he joined *Resolution*, another from injuries sustained when he fell down an open hatch and two who drowned. Not a single death from scurvy!

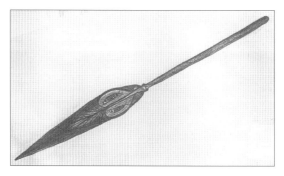

Polynesian paddle

18. Cook Agrees to Command the Third Voyage

July 1775 to January 1776

Adventure had been home for a year, before *Resolution* arrived. Naturally, Furneaux by arriving first, had stolen some of the limelight. Furthermore, he had brought back a real live South-Sea islander. Cook had tried to discourage Furneaux from taking Omai to England, he did not rate Omai's intelligence highly enough to hope for the help Tupia had given. He feared that, once the novelty had worn off, Omai would end up in some London slum, or as a curiosity in a fairground.

Omai was a bit of a braggart and a weaver of fanciful tales. He clearly expected to return from England with his prestige greatly enhanced and talked of taking guns and gun powder to arm his countrymen on Raiatea to enable them to free themselves of the yoke imposed by the Chief of the neighbouring island of Bora Bora.

Furneaux presented Omai to the Admiralty, Joseph Banks, as the leading authority on Polynesia, was summoned to interpret. He was in his element and quickly took Omai off Furneaux's hands. Within days, he presented Omai to the King and this was followed by a whirl of social engagements. Omai was now dressed in the latest London fashions, acquiring the finest manners and learning to gamble. People saw him as the embodiment of the 'noble savage.' He was invited by Dr Johnson and was praised by James Boswell, and others, for 'his impeccable manners.' Sir Joshua Reynolds, Nathaniel Dance and William Parry all painted his portrait. Lord Sandwich took him to live on his estate, where Omai spent a few months teaching the Polynesian language to his host and to Joseph Banks.

Banks and Omai were soon seen as the centre of interest of the voyage that *Adventure* had just completed, even though Banks had not been part of it. The newspapers were full of tales of the South Seas, of Banks and Omai and of *Adventure*'s voyage and, what they didn't know, they were happy to invent. The seamen had been upstaged once again.

Captain Furneaux's 'exploits' did not cut much ice with the Admiralty, however. They held him partly responsible for the incident during which his men had been killed and eaten and, although they respected the fact that after the separation from Cook and *Resolution*, Furneaux, instead of hurrying home directly from Queen Charlotte's Sound, had stuck to the southern course, passing several hundred miles

south of South America, then continuing at the high latitudes until he was due south of Cape Town – only then making his way back to England. This redeemed him somewhat in the eyes of the Admiralty, but not enough to earn him a promotion. He was given a year's leave, then he was put in command of a frigate, *Siren*, shuttling up and down the North American coastline, as part of the war with the colonies.

When Cook had returned from his first voyage, he had, in accordance with his orders, handed not only his own logs and journals to the Admiralty, but those of his officers and petty officers too. Once the Admiralty had gone through all this information, the next task was to publish it. To this end, they had engaged the services of Dr John Hawkesworth, a freelance writer of modest beginnings, but by this time, modesty had taken a back seat. He was entrusted, not only with the publication of the literature concerning Cook's voyage, but also that of the voyages of Byron, Carteret and Wallis. It was important to publish these narratives, as it was part of the process of laying claim to the territories discovered.

Hawkesworth was on to a good thing. Just how good is illustrated by the payment of £6,000 he received from a printer to secure the right to print the accounts of the voyages. £6,000 would be a useful sum today, but in 1773, when a good farmhand might be paid £3 a year with board and lodging, it was a sum beyond most people's wildest dreams.

Hawkesworth had been handed the hackwriter's dream on a plate. He was given all this precious material and told to use it to produce a narrative that would hold the interest of the ordinary reader, and that's what he set out to do. He was a firm believer in the maxim: 'Don't spoil a good tale by sticking to the truth.' He embellished and invented to his heart's content. His works were published in late 1773, while Cook was still in New Zealand, and they were an instant success. They were translated into German and French. The first English edition was soon sold out and demand was such, that it was reprinted a few months later.

When Cook had reached the Cape, he had been handed a copy of 'his' journal and had been horrified. Not only did it contain appalling mistakes and unduly flattering references to Joseph Banks, with sections attributed to Cook which had really been written by Banks. It also claimed that Cook had read and approved this version of

Opposite: *Portrait of Omai, the young South-Sea islander who was to become the toast of the town, upon his arrival in London with Tobias Furneaux.*

events. There was nothing Cook could do, but he swore to himself that he would keep a strict control of his journals for the present voyage.

Cook, Byron, Carteret and Wallis were all horrified and embarrassed by the way their accounts had been twisted out of shape. Hawkesworth had included unjustified and highly critical comments in his book on the behaviour of the people in power in St Helena, attributing them to Cook. When the latter had stopped there in May, and gone to pay his respects to the Governor, John Skottowe, who was none other but the son of Thomas Skottowe, of Great Ayton, who had been Cook's childhood benefactor, the Governor's wife and children took Cook to task about 'his' unfavourable remarks. They soon realised that he had had nothing to do with these comments and they lavished hospitality on him throughout his stay.

Three months before *Resolution* returned to England, hostilities had broken out in Massachusetts, between the Colonial Militia and the British troops. This was to be the start of the American War of Independence and it was one more thing that competed with Cook's return for newspaper space.

It also has to be said that the hardships and achievement of circumnavigating the Antarctic fell well short of the glamour of the Tahiti voyage and the tales Banks was able to spin. This too muted the reports of *Resolution*'s voyage. The press also found it difficult to report the fact that the Great Continent, which – in their eyes – the voyage had been in the business of finding, had not been found, in fact its non-existence had been proved. The papers were struggling to find a way to make Cook's voyage seem 'newsworthy' to their readers, now that the long-cherished dream of the Southern Continent had been proved to be nothing but a colourful myth.

Once more, the Admiralty made up for the lack of public interest. A week after his arrival Cook was presented at Court and, not long after, he was promoted to Post Captain. This time it was a promotion worthy of the man and his achievements. If an officer of this rank continued to serve in a competent and praiseworthy manner, he was almost assured of continued promotion and, if he stayed the course, he had a good chance of reaching the top, or very close to the top, of his profession.

Opposite: *Sir Joseph Banks painted by Benjamin West*, ca *1773. In this portrait Banks wears a Maori cloak and is surrounded by a variety of artefacts brought back during the* Endeavour *voyage. These objects were highly sought after by collectors and, during Cook's subsequent voyages, the crew were eager to bring back as many of them as they could lay their hands on.*

In the autumn, Cook was nominated to become a Fellow of the Royal Society. It was usual for three, or possibly as many as six Fellows to sign the proposal. In Cook's case, no less than twenty-five Fellows signed. Wales, the astronomer who had travelled with Cook, was also nominated, and both men were officially accepted as Fellows in early 1776.

The Copley Gold Medal was awarded annually by the Royal Society to an individual who had made outstanding contributions to science. In July 1776 it was awarded to James Cook. This, the Society's highest award, was given to Cook in recognition of his work in banishing scurvy. Cook had been told before he left England on his third voyage, that the medal was to be given him. In his absence it was presented to his wife. In the same year he published a paper on the health of seamen on long sea voyages in the *Society's Philosophical Transactions*. It was a summary of the way he had beaten scurvy in both of his Pacific voyages. He continued to advocate sauerkraut as part of the diet, although we now know that it was far from being as valuable as he believed. However, the concentrated citrus juice that was administered as a cure for those who were showing the early signs of scurvy, was most effective. But his insistence on the importance of providing fresh food at every opportunity, was the real breakthrough.

Cook also strongly recommended that ships should always renew their stock of fresh water, whenever the opportunity arose. Water, taken from the mouth of rivers, was often of doubtful purity and would deteriorate rapidly when stored in casks. He was also adamant that the crews' quarters should be kept clean, dry and well ventilated. These measures seem obvious to us, but they were a real novelty in his day. He provided his men with the best of available warm clothing and insisted that it should be washed and kept dry, although this must have been very difficult in bad weather.

Whenever possible, Cook had his crew working a three-watch system which meant that, barring emergencies, each man would work a four-hour shift and then have eight hours off. The watch that fell between 1600 and 2000 would be split into two, two-hour watches – the dog watches. This was a considerable improvement on the two-watch system, which required a man to work four hours on and four hours off, day and night, week after week. The modified lookout routine put in place by Cook, whenever it was very cold, required the lookout to go to the masthead every thirty minutes, look round, report and come down to the deck again, rather than having to perch there for four hours at a time.

In his reports to the Admiralty, Cook made little of his epic circumnavigation of the Antarctic, but focused on the health of his crew and the means that allowed him to achieve this. It is difficult to decide which of Cook's contributions was the greatest, but when one thinks of the countless lives his measures must have saved, it is tempting to believe that this was his true triumph.

There was talk of a third voyage to the Pacific, in a refitted *Resolution*, under the command of Charles Clerke, its prime purpose being the return of Omai to Tahiti. Cook was happy with this idea, as he was all in favour of the return of Omai to the islands, but it was not a task that he wanted to undertake himself. To him, it lacked a sense of purpose.

Cook made a surprising move at this point. He asked to be put in command of His Majesty's Royal Hospital for Seamen, in Greenwich. This post was something of a sinecure for a naval officer who felt that the time had come to be 'put out to grass'. It was a task that held little or no challenge, even the routine work would be done by lieutenants. But it also was a prestigious appointment, that would ensure him a comfortable income for his retirement and security for his family. It is possible that Cook may have felt weary after two such taxing voyages, his state of health was probably far worse than he let on, and he may well have been anxious of how little he had seen of his beloved wife and their growing children. It seems that even *he* was perhaps a little uncertain of the suitability of the appointment, as he asked that it should not block his acceptance, should anything of sufficient interest arise.

A week after the Greenwich appointment, Cook wrote to his old friend, John Walker, of Whitby. His letter points to his mixed feelings: 'A few months ago the whole Southern hemisphere was hardly big enough for me and now I am going to be confined within the limits of Greenwich Hospital, which are far too small for an active mind like mine, I must however confess it is a fine retreat and a pretty income [£230 per year and residence], but whether I can bring my self to like ease and retirement, time will shew...'

Much of Cook's time was spent preparing the journal of his second voyage for publication... there was no way he was going to let Hawkesworth get his hands on it this time! Things did not go smoothly for him. Cook learned that John Marra, a Gunner's Mate aboard *Resolution*, had kept his own journal and was preparing it for publication. On investigation, Cook decided that it did not offer a threat to his own work and did not obstruct its publication.

When John Reinhold Forster had agreed to sail with Cook, he had asked the Admiralty to allow him to publish his own account of the voyage. The details of the agreement were somewhat vague and, after much argument and some aggravation, it was agreed that he and Cook should consider a joint venture, with Forster tackling the scientific side, and Cook writing the narrative of the voyage. Things proceeded on this basis, throughout the autumn of 1775 and the winter of 1776.

Cook had enlisted the help of a clergyman, the Reverend John Douglas, Canon of Windsor, who corrected his grammar and spelling, but he also sanitized the material by censoring references to the nakedness of the people of some of the islands Cook had visited. Of course, any suggestion of sexual activity was excluded.

It seems that part of the agreement was that both men should submit their work to Lord Sandwich who decided that Forster's work overlapped Cook's in many places and should be edited to eliminate duplication. Forster, a proud and obstinate man, would not accept this, it was too much for him to acknowledge that his work should be tampered with. The dispute dragged on for some time until, at last, Forster withdrew from the joint venture altogether.

Cook pressed on and, in 1777, his book was published under the somewhat cumbersome title of *A Voyage Towards the South Pole and Round the World*. Happily for Cook, it was an instant success and it sold out on the first day it was offered for sale. A reprint appeared a few months later. The book was translated into French in 1778 and two more editions in English were produced, one in 1779 and the other in 1784.

Meanwhile, Forster had been hard at work, producing the first of his scientific works in Latin, *Characteres Generum Plantarum*, which gave details of ninety-four new plant species.

Cook tried to block the publication of this work on the grounds that it was an infringement of his sole right to publish the first account of the voyage. The attempt failed, as Forster was able to satisfy Lord Sandwich that a botany treatise, written in Latin, was totally separate from Cook's account of the voyage.

Three years later, Forster published another book, under the title of *Observations Made During a Voyage Round the World, on Physical Geography, Natural History and Ethnic Philosophy*. More than half of the book was devoted to what we would call anthropology today, and it had a considerable impact on what was then an infant science. For all his shortcomings, Forster was a fine scientist.

The agreement Forster had signed with the Admiralty prevented him from writing about the voyage, as opposed to writing on scientific matters arising from the voyage, but it dawned on him that there was a simple way to get around this agreement. His son, George Forster, had not been party to the agreement and could, therefore, publish what he liked.

In fairness to Cook, it has to be said that Forster was a difficult man who had alienated just about every officer aboard *Resolution*. Erudite, and undoubtedly an asset to the voyage from the scientific point of view, he was also described as 'dogmatic, humourless, suspicious, censorious, pretentious, contentious, demanding'. Such attributes aired in the cramped conditions of a sailing ship, must have been exasperating. He was hated by the crew and in constant threat of being thrown overboard. He so angered Charles Clerke that the latter threatened him with arrest. On one occasion, he was punched to the deck by the Master's Mate. Before joining *Resolution*, Forster had tried his hand at teaching and had offended his various employers so thoroughly, that he was repeatedly dismissed from his post. Luckily, his son made up for a lot. He was an excellent naturalist, a talented artist and a linguist.

When the Forsters' second book came out, Cook's champions attempted to denigrate it, but in the end they gave up. All parties concerned, including Cook himself, saw the writing of the books as an excellent money-making activity, as well as a way of advancing knowledge, and the market was obviously big enough for them all to benefit.

In the meantime, the search for the Great Southern Continent had been replaced by that of the Northwest Passage. The latter of course is not a myth but it would take another 150 years before it was discovered. However, by the time Cook returned in 1775, a voyage of discovery seeking the sea route to Asia was being planned. The Admiralty Lords obtained an extension of the 1745 Act which offered the massive sum of £20,000 for the discovery of the passage. The job seemed tailor-made for Cook, but of course, he was now heading for his comfortable sinecure at Greenwich.

In January 1776, according to Cook's first biographer, Andrew Kippis, who wrote this account twelve years later, Cook was invited to dine with his friends and patrons: Lord Sandwich, Sir Hugh Palliser and Philip Stephens of the Admiralty. It was Stephens who had suggested to Palliser that Cook should command the first voyage, aboard *Endeavour*. The farm worker's son had grown in stature with each voyage and was now perfectly at ease in the company of such high-ranking men. They were indeed friends of his in the truest sense. Palliser had nurtured Cook in his early years, recognising the potential that was there and continued his patronage throughout Cook's career. Lord Sandwich, who was in a position of power at the

Admiralty, had watched over Cook and had eased his path for almost as long as Palliser. Cook had been invited to dinner, ostensibly to advise on the preparations for another voyage to the Pacific, under the command of Charles Clerke. The three gentlemen probably wanted Cook to command the voyage, but did not know how to ask. Cook was 47 by then, advancing age in his day, he had been given a prestigious appointment which would give him a comfortable retirement and a chance to enjoy family life. Yet, before the evening was over Cook had offered to command the forthcoming voyage and his offer had been unanimously accepted.

Lord Sandwich immediately went to the King, who gave his instant and enthusiastic approval. Cook was warmly congratulated by the Royal Society and by the Navy. He found himself the hero of the day.

Seal

19. Preparations for the Third Voyage

January 1776 to July 1776

Resolution had been earmarked for the next Pacific expedition and her refit had been in hand for some time, before Cook had volunteered his services for a third voyage. As a result, a great deal of work had already been done on the ship, which Cook had not been been there to oversee.

Adventure was away on another task and the Admiralty Surveyors found another Whitby-built ship, *Diligence*, to take her place. She was rather smaller than *Resolution*: 295 tons against *Resolution*'s 426. Her name was changed to *Discovery* and her conversion was put in hand.

When Cook had volunteered to take on the command of the expedition he had in fact deprived Captain Clerke of that position. He did not know if Charles Clerke knew this, but the friendship between the two men and Clerke's character were such as to exclude any feelings of resentment. Clerke was appointed to command *Discovery* and to be second in command of the expedition. If he lacked ability in surveying and drawing, he was very good with his men, who held him in high esteem. Clerke had been with Cook aboard *Endeavour* and *Resolution*, as well as with Byron on his circumnavigation. He was probably one of the most experienced seamen of his day. In short, he was the sort of man you would want around, when the going got tough.

Cook chose John Gore, an American, as his First Lieutenant, another man of considerable experience. He had crossed the Pacific with Byron and, again under the command of Clerke, aboard *Dolphin*. He lacked Cook's flair, but he was a competent and steady seaman. Importantly, he and Cook were good friends. James King, young, studious and recommended by Palliser was the Second Lieutenant. John Williamson was Third Lieutenant. William Anderson, the Surgeon, who was also a gifted natural historian and a linguist, completed the after guard.

Lord Sandwich suggested William Bligh for the important post of Master. Like Cook, Bligh had come up through the hawsepipe, eventually rising to commissioned rank through sheer ability. At the age of twenty-two he already had a reputation as an exceptionally capable navigator and cartographer. He was a first-class seaman and though not the monster portrayed by Hollywood, he was prone to violence and unduly aware of his humble origins, which did not make him an easy person to have around.

Molesworth Phillips was appointed Lieutenant of Marines aboard *Resolution* and was to prove himself an idle and incompetent officer. Samuel Gibson was with Cook on his first and second voyages. He was the Marine Private who had tried to run away with a Tahitian girl on the first voyage, been given a second chance by Cook, who promoted him to Corporal during the second voyage. Now he was the Marine Sergeant and a very good one by all accounts. Cook had a soft spot for him and, needless to say, Gibson worshipped him. His ability and loyalty undoubtedly helped to counter the inadequacies of Lieutenant Phillips.

On board *Discovery*, Captain Clerke had James Burney as his First Lieutenant. William Bayly was again on board as the ship's astronomer. William Ellis, a friend of Banks and – strange for his day – Cambridge-educated, was the Surgeon's Mate. He was also a brilliant painter. David Nelson had been appointed at Banks's special recommendation. He worked at the Royal Botanical Gardens in Kew and it was his task to collect specimens of the flora during the voyage.

On this third voyage, there were to be no supernumeraries, apart from Omai and the voyage artist, John Webber. Recommended by Daniel Solander, Webber turned out to be a skilled and prolific illustrator. There were no official scientists on board either. It may be that after his experiences with Banks and Forster senior, Cook was weary of the breed.

Omai was to be returned to Tahiti aboard *Resolution*, together with an astonishing amount of baggage which included a coat of mail and a suit of armour, not to mention a bed, table and chairs. Cook made no difficulty over the amount of baggage, but he did jib at the muskets and shot Omai insisted on taking with him. However, the Admiralty Board overruled Cook on this last point and Omai was allowed to sail with his arsenal.

Cook received two sets of orders: the open orders instructed him to go to New Zealand via the Cape of Good Hope, searching en route for the islands the French had claimed to have discovered and fixing the position of any he might find. From New Zealand he was required to return Omai to Tahiti and then sail to the west coast of North America at about latitude 45º north. From there Cook was to proceed to the north, to latitude 65º and advised to arrive there not later than the following June. He was instructed not to spend time investigating inlets and rivers, until he reached this latitude.

The sealed orders specifically mentioned the search for the Northwest Passage, an objective of which the public had been kept ignorant. According to the schedule put forward by the Admiralty, Cook should be able to start on his search for the Northwest Passage around the beginning of 1777. To make the most of the short spring and summer seasons, Cook was to head for New Albion (Drake's discovery) on the American coast, sailing north from a latitude of around 45º to 65º, the latter being just short of the Arctic Circle. The consensus was that if a Northwest Passage existed, it was likely to be in that area. Should a likely inlet be discovered, but was found to be too shallow to allow *Resolution* or *Discovery* to proceed, Cook had on board the prefabricated frames of two smaller ships, which could be set up, planked, decked and rigged. The sealed orders also contained political advice on how to avoid giving offence to the French or the Spanish. Shortly before Cook left on his third voyage Britain found herself at war with America, Spain and France, All three countries told their naval forces that Cook was not to be treated as a combatant, but to be given any assistance he might require. Although he was sponsored by the British government, his work was recognised to be of international importance.

Richard Pickersgill, a veteran of the two previous voyages, was not part of the third. Instead, he was sent to Baffin Bay on an expedition organised by the Admiralty, which involved conducting a search for a waterway that led inland from the Atlantic, with the hope of meeting up with Cook. This expedition was to be a complete failure and poor Pickersgill took to drink and was dismissed from the Navy.

Resolution's total complement numbered 112 and *Discovery*'s was seventy, but this was not all. King George, known in some quarters as Farmer George, was keen to populate the Pacific islands with European livestock. *Resolution* had to find deck-space for a bull and two cows with their calves, a large number of sheep, pigs, goats, chickens, rabbits and a peahen and peacock. In addition to the food and water for these creatures, there was a considerable quantity of a wide variety of seeds to be stowed in dry and rat-proof conditions. All this was in addition to the normal ship's stores such as timber, rope, sail cloth, gunpowder and shot, not to mention items like 700 pairs of trousers and 800 pairs of shoes and three-and-a-half tons of sugar.

The condition of *Resolution* left much to be desired. She had had a tough time on her last voyage with Cook. The many thousands of miles she had sailed in the Pacific, the Atlantic and in the Antarctic Oceans, had taken their toll. The Deptford dockyards were known for their slackness and corruption, especially if a close watch was not kept on their activities. In the past, Cook had been able to watch over the work as it was done, but, this time he saw nothing of the first six months of the refit. He was then caught in such a whirlwind of social engagements that his usual painstaking attention to details relaxed somewhat and, in any case, the shoddy work had probably already been carried out.

As the months passed, Cook discovered that he had indeed become the toast of the town. His portrait was painted by Nathaniel Dance at Banks's instigation and his expense. He was invited to dinner at the House of Commons, he dined with the other Fellows of the Royal Society, had several meetings with James Boswell, who was much taken with Cook, even asking to be taken on his third voyage (he was dissuaded from this plan). Whenever possible, Cook took his wife to these grand occasions. She took it all in her stride. The same Boswell described the couple affectionately: as a 'grave steady man, and his wife, a decent, plump Englishwoman.'

Meanwhile, vital tasks such as the caulking and copper sheathing of the hull and the caulking of the deck seams which would keep the ship dry, had been carried out, as well as the examination and repair, or replacement, of the masts and spars. It was not always easy to check on the quality of the workmanship. Many of these items could not be examined, without undoing much of what had been done already. Cook was reluctant to do that, as time was not on his side. This failure was to cause him a number of problems in the not too distant future. Partly due to lack of time and partly because Cook felt that little structural work had been carried out on *Resolution*, there were no sea trial of the vessel before they sailed. The dice seemed indeed loaded against the voyage from the start. It could be argued, that if Cook had not been forced to return to Hawaii for repairs to his ship, he may well have lived to see the end of the passage and there was little doubt that the need for repairs could be traced back to the shoddy work of the Deptford dockyards.

As departure neared, Cook displayed more concern about the difficulties and dangers of the voyage that lay ahead , than he had ever shown before. The main thrust of his orders was to seek a Northwest Passage from the North Pacific to the North Atlantic. The voyage would undoubtedly be long, hazardous and demanding – physically and mentally. Yet, all the tasks he had carried out for the Navy in the past had come with their own special element of danger and Cook had never seemed daunted by it. In letters he wrote to friends at this time, he betrayed a nervous anxiety and made several references to the dangers inherent to the forthcoming voyage, and this was out of character.

I wonder if his health had something to do with his low spirits. He was approaching fifty and that was a good age for his day, although there were senior officers that were, or had been, fighting battles successfully at seventy. He had been very seriously

Opposite: Sir Hugh Palliser *by George Dance. Sir Hugh was a life-long friend and supporter of Cook, who named one of his sons after him.*

ill on his last voyage and it is very likely that the malady he had suffered from had not cleared up completely, and may indeed have affected his health permanently.

The residual bad health, combined with the onerous nature of his work, must have caused him a good deal of tension. His actions and his decisions could affect the well-being, or even the lives, of more than a hundred men. He would have been an unfeeling person for this not to have weighed heavily on his mind at times of crisis. And, if we can be sure of anything, is that Cook, whatever else he may have been, cared for those under his command. He may have been on friendly terms with his officers but, nevertheless, the final responsibility was his and command is a lonely and solitary business.

His concern about his future also led him to worry about the welfare of his wife and family, should he die at sea. He confided these worries to Lord Sandwich, who assured him that official provision would be made for Elizabeth, should something untoward happen to him. At least, the spectre of his wife and family being consigned to the less than tender care of the workhouse, was banished from Cook's thoughts. Meanwhile, Elizabeth gave birth to another boy, Hugh – named after Sir Hugh Palliser – on 23rd May 1776.

Coping with the preparations for the voyage, his many social obligations and with the distraction of Elizabeth's pregnancy, Cook had also striven to complete the preparation of his journal for publication, which he managed to do before he left. It was a massive task that would produce two large volumes. It was not surprising that Palliser found Cook tired and worn, when he went to visit the new-born babe.

There were several other social functions to attend, before Cook could return to his home in Mile End and spend a little time with his family. Early on 24th June he said farewell to Elizabeth and his children, boarded his carriage and made his way to the Thames, where *Resolution* was anchored.

At the end of June Cook anchored in Plymouth Sound to take on a final load of fresh food for the crew, and water and fodder for the vast array of farmyard animals they were taking with them. The contingent of Royal Marines came aboard.

Unfortunately, fate was not yet finished with this expedition. Charles Clerke was on board *Discovery* and ready to leave, when he received a summons from the King's Bench to answer for the debts of his improvident brother, who had fled the country. Clerke had stood as a surety for him. Lieutenant Burney was authorised to take charge of *Discovery* and sail her to Plymouth, while the unfortunate Clerke was thrown into a debtors' prison in St George's Fields in London, where he stayed, despite the best efforts of his influential friends, who included Lord Sandwich and the Speaker of the House of Commons.

Cook chafed at the bit in Plymouth. Ten days after his arrival, and despairing of Clerke's arrival, he wrote to him, saying that he was going on ahead and instructing him to sail for Cape Town, as soon as he was released from prison. On 12th July 1776, eight days after America's Declaration of Independence, *Resolution* finally left for the Cape. It was to be three more weeks before Charles Clerke was freed. Once out of prison, he made his way down to Plymouth, stayed one day there and then stood out to sea, intent on catching up with Cook.

Tuberculosis was rife in the London slums and prisons and, although he was not yet aware of it, the unfortunate Charlie Clerke had been infected.

Frozen sea

20. England to Moorea via Cape Town

July 1776 to August 1777

The passage to Table Bay, Cape Town, was an easy one. For most of the passage they had fair weather, but there were some periods of heavy rain which enabled them to fill some of their water casks. Unfortunately, the rain also quickly revealed the poor workmanship of the dockyard men who had caulked the deck seams. Everyone in the ship suffered from the water, that poured from the deck and topside seams, onto their bedding and much else besides. The crew were employed in re-caulking the deck seams, whenever they could be spared from other duties. The topside seams had to be tackled from inside the ship, as it was too rough to attempt to caulk them from outside. This is never satisfactory, as the planks are made as to leave a slender V-shaped groove between them to accept the caulking material, with the base of the groove on the inside of the ship. The sails in the sail room were soaked and they rotted, before the weather cleared up sufficiently to allow them to be dried out.

Resolution called in at Santa Cruz, in Tenerife, where they picked up some wine and some fodder for the livestock. Cook also took the opportunity to check the accuracy of the chronometers and to satisfy himself as to the longitude of the island.

A week out of Santa Cruz, *Resolution* was approaching the Cape Verde Islands when, at 2300, the Surgeon came on deck for a breath of fresh air and to watch the little island of Bonavista, as they sailed past it. He thought he could see a line of breakers ahead and was contemplating calling Captain Cook, whose watch it was, when the latter came on deck. As soon as he arrived, he spotted the line of breakers and ordered the helmsman to steer hard to starboard, calling to the duty watch to trim the sails in accordance with the new course. The urgency of the orders and the flurry of activity brought the off-duty watch out of their hammocks to lend a hand to the men on deck. The ship was a little more than a mile from the breakers, before the situation was brought under control.

Bligh was horrified that the Captain had allowed them to come so close to disaster when they had the whole of the Atlantic in which to sail. Cook was just as horrified and, doubtless, many of the crew were alarmed at their close brush with death. Their faith in their Commander's superb navigational skills, that had kept his men safe in so many difficult situations, was badly shaken.

Was this perhaps an indication that Cook's abilities were waning? It is possible that he was tired, closed his eyes and was asleep before he realised it, something that has happened to us all, I am sure. But why lay off a course that takes the ship anywhere near such a hazard?

There was more of the same to come. On the following day, in broad daylight, Cook mistook one of the Cape Verde islands and placed his vessel uncomfortably close to another set of breakers. He had to alter course to windward to avoid them. For a second time in twenty-four hours, emergency action had to be taken to avoid disaster. On each occasion, *Resolution* was forced to sail uncomfortably close to a lee shore to escape.

After leaving Tenerife, the wind made it necessary to sail in the direction of Brazil, until it served to take them to the South African coast. By the third week of October 1776 *Resolution* was anchored in Cape Town.

The livestock was put ashore and, one night, some dogs got into the sheep's pen. Four of the sheep were killed and the rest scattered. Some were recovered eventually and a number of local ones were purchased to replace the missing. As if there were not enough animals for the crew to care for and feed, Cook bought some horses!

The sheep incident upset Cook to the point of raging anger and he made a highly critical entry about this in his log, also about the treatment meted out to a French vessel that had been driven ashore and plundered by the locals, it appears, with the connivance of the police. These unrelated incidents were not that serious in themselves, but Cook's reaction to them was perplexing. *Discovery* arrived on 10th November. She had been caught in an exceptionally severe gale that had driven her away from the African shore. *Resolution* had felt the strength of the same gale in Table Bay and she appeared to have been the only vessel in that huge bay not to have dragged her anchor.

The caulking work carried out on *Discovery* back in England, had been done just as badly as the one on *Resolution* and the ship's company had suffered in the same way of the wet and the cold. As soon as the re-caulking of *Resolution* was finished, Cook sent his men over to help with the work on *Discovery*.

The astronomer William Bayly's painstaking celestial observations had established that the chronometer they had on board was losing a little more than two seconds a day, but as the discrepancy was consistent, it was easy to allow for in their calculations.

At the end of November both ships were ready for sea and they left Table Bay on 1st of December. This was the first time they had sailed in company. Twelve days out, they sighted a small group of rocky islets, that had first been reported by Marion du Fresne in 1772. The depth of water around them was such, that they could sail with safety to within half-a-mile of the shore. The islands were totally barren, there was nothing to be seen but rock.

On 24th December they fell in with Kerguelen's Land which Cook renamed Island of Desolation. The Kerguelens are a group of three hundred volcanic islands, that lie more or less halfway between Africa and Australia. Cook saw them as bare and inhospitable, but there is a snug bay on the northeast coast, which both ships entered and anchored in. In view of the date, Cook named it Christmas Bay. Once all was secured the men were allowed ashore to stretch their legs. They gathered anti-scorbutic cabbage that was one of the few plants on the island and killed many penguins and seals for meat. Cook and Bligh worked together for a week, surveying and plotting the lay of the land.

By New Year's Day 1776 they were on their way, with Tasmania as their next stop. Cook's instructions from the Admiralty had envisaged that they would sail directly from Cape Town to Tahiti, a distance of at least 11,000 miles. To cover such an enormous distance, without taking in fresh food and water for the men, wood for the galley and fodder for the animals, was well nigh impossible. Furthermore, it made no allowance for possible repairs to the ship. Even forgetting the delayed departure from England, the Admiralty's instructions were clearly unrealistic.

The passage across the Indian Ocean was made in a latitude of approximately 48º south, a big improvement on their earlier crossing which was made mostly in the latitude of 60º south. But although the weather was warmer, the seas were rough enough at times to reveal more of the shoddy work carried out by the English dockyards. Both vessels still leaked badly, despite the best efforts of the crew to remedy the substandard caulking and, every time it rained, or when the wind piped up, everyone had to endure wet conditions.

There were long periods of dense fog, that made it difficult for *Resolution* and *Discovery* to stay in touch. At times it was so thick that it was impossible to see from one end of the ship to the other.

Three weeks out from Kerguelen, they were hit by a vicious squall that toppled *Resolution*'s fore top mast and her top gallant mast. Mercifully, the top mast did not fall to the deck, but swung about, supported only by the halyards and some of the forestays. Seamen hastened to their station and all was a flurry of disciplined haste, the men swarming up the ratlines, until they could reach the broken section of the mast, take control of it and, eventually, lower it to the deck. This kind of incident is not that rare at sea, but in this case, the masts snapped, probably because the vessel was carrying too much sail. This was the kind of mistake which Cook had never been known to make. He may have been cutting corners in the vain attempt to catch up with the timetable, but, those men who had been with him on previous voyages, were full of consternation. They had always believed that Cook would be as unsparing of ships and men as he was of himself, but that he was never reckless. His usual prudence and inspired seamanship seemed to be deserting him.

A day was lost clearing away the mess and stepping a new top mast. The top gallant would have to wait until the shipwrights could make a new one, they had no spare on board. Cook did not discuss the incident with his officers and made for the Tasmanian coast. In his journal he refers to the incident as if it had been a lucky opportunity 'to carry into execution a design [he] had formed of putting into Adventure Bay to get a little wood and some grass for [the] cattle.'

On 26th January 1777, they arrived in the bay, with its beautiful and well sheltered anchorage, which Furneaux had named Adventure Bay. The spot seemed to have been designed to refresh travel-weary crews. As soon as the two vessels were anchored, boats were lowered and parties sent ashore to collect fodder, wood and fresh water. The botanists and Bayly, the astronomer, had gone with the first of the men and Cook was not far behind. He left some of the pigs in the woods, hoping that they might not be caught by the Aborigines, before they had a chance to breed. (Apparently, this was not to be.)

Mindful of *Adventure*'s previous troubles with the natives of Tasmania, Cook put four marines ashore with the working parties. Far from acting as a guard, the marines purloined the men's alcohol ration and proceeded to drink themselves legless. They were hauled aboard unconscious and the next morning, they were given a dozen lashes apiece.

William Bligh was skilled in drawing and watercolour painting and, wherever they stopped he would record the scene. Later on, his pictures were copied by engravers and used to illustrate the pilot books of the day. Bligh also started work on a chart of the bay. Bayly was setting up his observatory. These activities were taking place under the watchful eye of the locals. Omai and some of the men tried to communicate with the Tasmanian Aborigines, but without much success. To the sailors' consternation, the women were no more disposed to pleasantries than their men.

Suddenly, Cook seemed to regret having come to the bay at all and announced that they would leave, as soon as the wind was favourable. The ship's company was getting used to their commander's shifting moods and indecision. On 30th January Cook ordered that the anchors be raised, and they were on their way to New Zealand and Queen Charlotte's Sound.

The passage took eleven days, uneventful, except for the loss of one of these marines, who had been punished for being drunk, while on duty at their last stop. George Moddy managed to get himself thoroughly drunk again and fell overboard to his death.

Cook's landfall, Cape Farewell, was forty-five miles in error, which was well below the standard of his past performances. Was this one more indication that Cook's mental capacity was waning? The intention had been to enter Ship's Cove and collect water from the fine fresh-water stream. The shelter left something to be desired, but water was the attraction. When they had finished watering they would move on to Grass Cove. *Discovery* got in to the cove and anchored, but *Resolution* was suddenly becalmed. Her boats had to be put into the water to tow her into the anchorage. The date was 12th February 1777.

More than three years earlier Grass Cove had been the scene of the massacre of a number of Captain Furneaux's crew by the Maoris and, those men in the crews of *Resolution* and *Discovery*, who had been aboard *Adventure* when the attack occurred, were less than happy to be back again. The Maoris seemed wary. Cook ensured that the men were protected by the ship's guns or by parties of marines, whenever they went ashore.

This is the time when Cook and his men finally obtained the details of the tragedy that had befallen the crew of *Adventure*. A Maori, who had been Cook's closest friend, related the grim happenings, with Omai acting as interpreter. Early on that fateful morning, Rowe and his working party were eating their lunch ashore. Furneaux's black servant was keeping a lookout. Between them, the men only had a few guns. Before long, some Maoris approached and there was much drinking. A quarrel broke out, Rowe was known to dislike the Maoris whom he regarded as savages. Things worsened, there was some thievery and Rowe shot two men. Before the crew members had time to reload their guns, they were attacked by the two men's friends and bludgeoned to death.

Later, when the cutter was long overdue, Lieutenant James Burney, the Master, and ten marines were sent out in the launch to look for the party. They sailed the shores of Queen Charlotte's Sound all day, searching each cove and beach for some sign of the missing men. Whenever they landed, the few people who were about, fled, some of the men making aggressive gestures, before disappearing into the surrounding forest. As time wore on the young Lieutenant and his men became more and more concerned. The more they searched, the more they became convinced that some disaster had befallen their shipmates.

Eventually, they spotted a large catamaran and two men on the beach, close to it. When they sailed towards it, the men ran off. This only served to reinforce their suspicions. They landed on the beach, close to the catamaran, and examined its interior. To their dismay they found a piece of the cutter's gunwale and a number of shoes, that had unmistakably belonged to the missing crewmen.

Worse was to come, much worse. Nearby, they found some baskets and, when they opened them, to their horror, they found them filled with pieces of the missing men. Much of the flesh had been roasted and cooked. Fern root was spread amongst the human remains. Clearly cannibalism was intended.

They extended the area of their search and found further human remains and more shoes. They loaded all they had found into the launch. The day was nearly spent and darkness was coming, but they sailed on, round the small headland, into Grass Cove.

On the beach, one small and three large canoes were drawn up, but on the hillside rising from the beach, a great fire was burning, and the hillside swarmed with people. The atmosphere was as of a carnival. By now the horror experienced by the crew was replaced by a burning anger. The marines stood up in the launch and fired volley after volley into the assembled crowd, their only thought was that of vengeance, and who could blame them? The crowd of Maoris fled, screaming, into the surrounding forest.

Burney and the marines went ashore and the Master stayed to guard the launch. They found the bundles of grass cut by the sailors and ready to be loaded into the cutter. It was obvious that here was the spot where their shipmates had been slaughtered and butchered. The scene of carnage was such, that some of the men could barely talk about it, when they got back to *Adventure*.

It became apparent to the Master that the Maoris were organising themselves for an attack on the marine party and he ran to warn them. The men were besides themselves with rage and the Master had some difficulty in getting them to return to the launch but, eventually, he was able to convince them that they stood no chance against such vast numbers. Before they left, they smashed the canoes to prevent the Maoris from following them.

Darkness had now fallen and only served to increase the sense of menace felt by all in the launch, as she sailed back to *Adventure*. It was close to midnight before they reached the ship, but there was little sleep for anyone. The situation was discussed throughout the night. The consensus of opinion was for retribution, but Captain Furneaux's decision was that it would serve no useful purpose to attack the Maoris, a massacre and possible further losses among the crew, would not bring the dead men back.

The next day, the remains of the seamen were sewn into hammocks weighted with ballast and, as *Adventure* left that unhappy place, they were committed to the deep.

Now, over three years later, there were men aboard *Resolution* and *Discovery* – including James Burney – who remembered all too well the events of that terrible night. It is not difficult to imagine their feelings.

Cook met Kahura, the man who had allegedly killed Rowe. Instead of acceding to his men's passionate desire for punishment, Cook had the ship's artist paint a portrait of Kahura. He felt that there was little to be gained from belated punishment and that, in any case, Rowe and his men had been partly responsible for their fate. As was now his wont, he didn't bother to explain his reasoning to his companions and, to the men, it seemed as if Kahura were being recompensed for his crime. Their bitterness towards Cook was matched by the latter's sadness, now that the dream of his special friendship with the islanders had been shattered. He was also deeply disappointed by the behaviour of Furneaux's men and this may have contributed to his refusal to punish the Maori. Whether anything would have been gained from executing the culprit, is open to speculation, but Cook's disdain of his men's understandable grievance and sorrow, is what soured the relationship between them.

Cook seemed more and more out of touch with the feelings of the ship's company. An atmosphere of pent-up rage reigned aboard and, within a few days, the men's resentment expressed itself in small acts of rebellion such as theft. It was not long before open confrontation broke out. For the first time aboard one of Cook's vessels,

Opposite: *A private of the Marines. This armed force was established by the Admiralty in 1755 and its main role was to protect the seamen sent ashore on working parties and guard the ammunition store on the ship.*

the protective party of marines, the traditional buffer, between a potentially rebellious crew below deck, and the officers' quarters, looked as though it could be needed.

On 15th March 1777 Cook decided to respond to the small thefts committed by some of the sailors, by cutting the crew's salt ration down to two-thirds. Then the meat ration was also cut drastically, when some was stolen. With most of the animals on board dead, meat had become a highly precious commodity.

It turned into a battle of will, the men refusing to eat the reduced meat ration. Cook flew into one of his by now famous rages, he viewed their action as mutinous and ordered that the reduced ration would continue until the thieves confessed or were reported. The crew continued in their refusal to eat and no one confessed. In the end, and much to his exasperation, Cook had to give in. For him this was an unprecedented situation, he had never been at loggerheads with his crew before. Interestingly, he made no mention of the incident in his log, it was as if he wanted to forget it ever happened.

Cook lingered in Ship's Cove and to his officers he was uncharacteristically wasting time. They were by now well behind schedule, but Cook displayed a lack of drive, totally unlike himself. Fortunately, the fishing was good, there was fruit to be collected to add to their diet and the weather was beyond reproach. This was well and good, but the inactivity and inertia created an atmosphere of disquiet among the two ships' companies and this feeling was not helped by the men's mistrust of the Maoris, who were now camped close to the ships, in quite considerable numbers.

It must have been a great relief all round when, on 25th February 1777, they left Ship's Cove for Tahiti. It took them forty-eight hours to make the passage through Cook Strait and get free of the New Zealand coast, the wind coming from every quarter and, on several occasions, falling away to nothing.

Cook had built up a considerable fund of knowledge of the wind patterns in the South Pacific, but now all his experience and observations seemed to desert him. He wanted to sail the most direct route to Tahiti, partly to try to catch up with his schedule, but also because such a route would cover ground that he had not explored. But the winds lacked any real strength and, at times, dropped to a complete calm. Often they were right on the nose of the ship, all in all, a frustrating passage for Captain Cook and his crew.

It was to be nearly five weeks before they were in the vicinity of land. On 29th March, *Discovery* reported land in sight. It proved to be the most southerly point of the Cook Group, an atoll called Mangaia. Unfortunately, there was no channel through the reef to allow a ship of the size of either barque, to enter its central lagoon.

The ships passed through the numerous islands that made up the Southern Cook Group, but Cook was unable to obtain fresh food in any of them. Most of the atolls and islands were inaccessible, because of shoal waters or coral reefs. In any case, the inhabitants on one or two of the atolls showed such hostility, when the ships showed signs of attempting to land, that Cook decided to look elsewhere.

In an effort to get some fresh provisions, a need for which was becoming critical, Cook changed course to the Palmerston islands. At least, the wind favoured that course! On 13th April, he was rewarded: one of the Palmerston Group of islands came into sight. It was minute, not more than a few feet above sea level, but nevertheless, it provided a large amount of scurvy grass for the men, fodder for the remaining animals and fresh water.

Nomuku, in the Tongan islands, was to be their next stop. The Tongans told Cook of the existence of nearly a hundred nearby islands that were totally new to him, the largest being Fiji and Samoa. Oddly, Cook made no effort to visit any of these places, preferring to stay amongst the islands he was familiar with. After all, it was obvious that the schedule set out by the Admiralty was by now irretrievable, but Cook seemed unable to motivate himself.

At the end of April, they reached the Friendly islands, where they would spend eleven weeks. It was a relaxing time for the crews and it could have been idyllic,

Opposite: Poedooa, the Daughter of Oree, *painted by John Webber. This beautiful young girl was the daughter of the Chief of Ulietea, one of the Society islands. The painting was exhibited at the Royal Academy in 1785.*

except for their commander's increasingly odd behaviour. Cook was flying into terrifying rages one minute and throwing parties for the local chiefs the next.

The usual petty thievery committed by the locals, which Cook would have looked on with a degree of annoyed affection in earlier times, now turned into serious offences to be punished with utmost severity. A chief, who had been accused of theft, was flogged and was not released until his family brought a pig as compensation. Following that incident, Cook gloated 'that after this we were not troubled by thieves of rank.' In an age and in a profession not known for its indulgent treatment of rule-breakers, the crew and some of the officers were horrified by the treatment meted out, which, they felt, outweighed the crime. The floggings became increasingly brutal, some culprits receiving as many as sixty lashes – the Admiralty regulations stipulated that a sailor was not to be given more than twelve lashes a day – Cook no longer respected such restrictions in the case of Polynesians. He made scant mentions of these punishments in his journal, but others – such as Midshipman George Gilbert – described his hitherto much-admired commander's behaviour with increasing disquiet: 'Captain Cook punished in a manner rather unbecoming of a European, viz by cutting off their ears, firing at them with small shot, or ball, as they were swimming or paddling to the shore; and suffering [the sailors to] stick the boat hook into them.'

Charles Clerke had to obey his Commander's instruction to come down hard on petty thieves, but he carried it out in his own way. Instead of flogging the culprits to the point of tearing the flesh of their backs, he preferred humiliating them by shaving off the hair over half of their head, and throwing them overboard to make their embarrassed way back to the shore. He, along with the other officers, could see no possible reason for staying where they were, now that they had replenished their stores. But they restricted their questions and concerns to their journals. No one dared question Captain Cook openly, for fear of provoking his fury.

From the Tongan Group of islands the two vessels made their way to Waitepeha, at the eastern end of Tahiti, where they stayed for a week. They then moved to their old familiar anchorage in Matavai Bay. The livestock that had survived the voyage from England was in good shape and was set ashore. Captains Clerke and Cook rode the horses, to the delight of the natives, who had not seen horses and riders in action before. The horses were ridden each day until *Resolution* and *Discovery* left Matavai Bay, on their way to Moorea island.

Opposite: *The coastline of Moorea, photographed by the author.*

Moorea lies about twenty-five miles from Tahiti. The son of the king of the eastern part of Tahiti had been killed, while on Moorea, and this outrage called for revenge. Cook was asked for his support in the forthcoming armed conflict. He tactfully declined. Fortunately, Moorea sued for peace before Cook left Tahiti and the matter was settled amicably. Cook visited Moorea and spent two weeks on the island.

Omai and his possessions were taken to Huahine, where he wished to be put ashore, and the expedition then went on to Bora Bora and a few more islands Cook had visited on earlier voyages.

Years later, it would be learned that the Tongans, Cook's 'friendly' islanders had been so incensed by the latter's brutality, that a plot had been hatched to kill him and the crew. Luckily for Cook and his men, the conspirators were divided on the best way to carry out the plan and this saved the visitors' lives. Cook and his men remained unaware of the plot and sailed before the plans could be carried out. It is a measure of Cook's rapid loss of grip on reality that he remained convinced that the islanders loved him as before, even after he had ordered that the young son of a Tongan chief be thrown in chains for allegedly trying to steal his cat. It was this child's father, Finau, who had hitherto been very friendly, who had hatched the plot to murder Cook.

21. Tahiti to Cook's Death in Hawaii

December 1777 to February 1779

Making use of the easterly wind, *Resolution* and *Discovery* sailed to the north from Tahiti on 9th of December 1977. Fifteen days later the masthead lookout reported land ahead.

Lying just north of the equator was a small island and, since it was Christmas Eve, Cook named it Christmas Island (now Kiritimati Atoll). The two barques anchored in the lee of the island and parties were put ashore. They hoped to find fresh water, but there was none. All was not lost, however, the surrounding waters abounded in turtles, which provided a welcome supply of fresh meat for the crews of both vessels. By chance, while they were there, there was an eclipse of the sun, which enabled Cook to fix the position of the island with some precision and to check the accuracy of the remaining chronometer.

They were soon on their way north. Cook and Clerke had expected a long haul in the midst of an empty sea. In preparation, Clerke rationed the water and Cook issued the heavy weather gear to the crew. They were pleasantly surprised when, on 18th January 1778, halfway between Bora-Bora and America, they came upon two of the westernmost of the eight Hawaiian Islands: Niihau and Kauai. Cook named them the Sandwich Islands, modern-day Hawaii. Kauai has a volcano at its centre and the island looked fertile, with dense forests in its valleys, lush vegetation and numerous fresh mountain brooks. Cook knew that he had made an important discovery, no one having apparently even heard of the existence of these islands.

Canoes were soon alongside and the men in them looked exactly like the Tahitians, their language was similar but gentler. They were unarmed and they quite soon began to exchange fish and sweet potatoes for trinkets. As with the Tahitians when *Endeavour* first landed, the Europeans soon discovered that there was no knowledge of metallurgy in Hawaii. The inhabitants were just as eager for nails or any metal objects as the Tahitians and could also be just as light-fingered. Cook sent three armed boats under the command of Lieutenant Williamson, to look for a landing place and fresh water. He instructed Williamson that if he found it necessary to land

Opposite: *This crab – forty-five cm across – was photographed on Christmas Island.*

to step ashore, he should not allow more than one man at a time to leave the boat. As there were instances of venereal disease on board, Cook also gave the order that 'no women, on any account whatever, were to be admitted on board the ships.' A suitable landing place was found at Waimea, on the southwest coast of the island. Water was available, which was fortunate, as they were running low, and later, several tons were taken aboard.

The islanders were obviously in great awe of the ships and the men who inhabited them. Cook had a boat take him to the shore. He landed on the beach before a considerable number of people, who prostrated themselves before him and stayed that way, until he gestured to them to get up. What he did not know is that the Hawaiians regarded him as some kind of divine being. He invited some islanders aboard *Resolution* and noted in his journal: 'In the course of my several voyages, I have never before met with the natives of any place so much astonished as these people were upon entering a ship.' Their demeanour made it obvious that they had never received visits from Europeans before. Cook mistook this respect for meekness. He had no idea that he had stumbled upon one of the most sophisticated and martial societies of the Pacific region. All was well for the

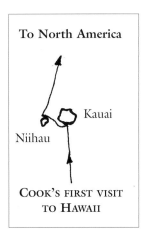

COOK'S FIRST VISIT TO HAWAII

time being, but it was obvious that the inhabitants of Hawaii were not likely to accept the lash, the way the Tongans and the Tahitians had done.

Cook was amazed by the way tiny children took to water, almost before they knew how to walk. Captain Cook himself could not swim – as a lot of seamen of that period – and the incredible empathy with water displayed by the islanders was a great surprise to him.

Very soon, an ugly incident took place, involving Lieutenant Williamson. He found himself surrounded by a crowd of islanders, intent on seizing metal objects from the small boat, he panicked and ordered his men to fire. One islander died on the spot.

The weather was uncertain and Waimea, where they had anchored, offered a rocky bottom and provided poor holding, even in a moderate wind. They shifted their berth a number of times, in an effort to find a more secure position. Finally they moved to the nearby island of Niihau, where they found the holding to be better. Shifting berth several times like this hampered the purchase of fresh food and the collection of water, which had been their prime reasons for stopping. As a result, they left on 2nd February, with less fresh food and water than Cook would have wished. *Resolution* and *Discovery* sailed on an approximately northeasterly course until 6th March, when they got their first sight of the mainland of North America.

The weather had been mixed en route, but now, it deteriorated quite badly, to the extent that Cook named a prominent headland Cape Foulweather. They suffered driving hail, sleet and snow and, at times, thick fog which reduced visibility to nothing. Weather like this, close to the coast, was dangerous enough, but the wind turned to the west and blew hard. Now the coast of America had become a lee shore. A lee shore is bad enough but combined with poor visibility, it turns into a nightmare situation. So bad was the weather, that it obliged Cook to seek safety in deep water, for much of the three weeks after he touched the coast of North America. The logs, of both Cook and Bligh, show their frustration at being driven so far offshore by the weather, as to make it impossible to do a running survey of the coast and off-lying waters.

Cook has been criticised, unfairly in my opinion, for failing to spot the Juan de Fuca Strait. It runs southeast for about fifty nautical miles, between the southern end of Vancouver Island and the mountainous headland that forms the western shore of the Strait. From a distance, and in the weather conditions described, the mountains of the western arm and the mountains of Vancouver Island could well appear to merge into one mountainous mass. Undoubtedly, driving along the coast road in an air-conditioned car, one would find it difficult to miss something as big as the Strait, but taking to the sea in a sailing vessel, without an engine and in exceedingly bad weather, one's perspective changes radically.

Three weeks after his initial landfall, Cook regained the coast and entered Nootka Sound, which is situated some fifty miles north of the Juan de Fuca Strait and about halfway along the west coast of Vancouver Island. Cook had originally named the Sound for King George, but the Admiralty changed its name to Nootka Sound, after Nootka Island which sits in a large bay on the western side of Vancouver Island.

After the severe weather both ships had encountered, they were in a battered condition and needed urgent repairs. Nootka Sound was perfect for their purpose: the large stands of coniferous trees would supply ample timber. All they needed was a safe anchorage, where they would be protected from the winds, should they pipe up again, and there were several to choose from.

As soon as the two barques were anchored in the Sound, four boats were lowered to the water. When they were rigged and ready to go, they sailed off to seek a better and more permanent anchorage for the ships. From the moment the vessels had anchored, they were surrounded with canoes, full of locals wanting to trade. Furs of all kinds formed the bulk of the articles on offer and, in exchange, they wanted knives, tools and metal. The inhabitants had probably been introduced to the value of metal by their contacts with Russian fur trappers and traders, and possibly by the Spanish seamen, who had sailed as far north as Alaska, although Cook was unaware of this at the time.

By noon, the boats were back, having found suitably sheltered spots for the vessels. But it was too late to move them that day and it was not until the following morning, 31st of March, that they shifted to their new berth. The water was deep and the bottom rocky, which ruled out anchoring, but it was possible to secure both vessels to the shore. The hands were set to caulking the leaky decks and topside yet again. This recurring problem really had to be beaten. To be constantly wet in tropical temperatures is unpleasant enough, but they were heading for the Arctic, and wet bedding and clothing in freezing temperatures could be fatal.

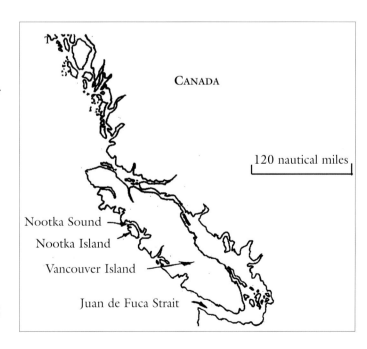

The masts also gave cause for concern. Whilst the shipwrights went ashore to cut the required timber, the seamen lowered the fore topmast to the deck and removed the rigging, so that it could be lifted out of the ship and taken ashore. Initially, it was thought that one of the bibbs (the huge wooden brackets at the top of the mast that support the platform) was rotted and needed to be replaced. But when the shipwrights started work, they found that the damage to the structure of the mast was even worse than they had anticipated. It was yet another evidence of the botched-up job carried out by the dockyard at home. The poor workmanship had allowed the rain to get into the timber. It is a strange thing, but sea water doesn't rot timber, but let fresh water get trapped, and you have a recipe for disaster. This is exactly what had happened.

The mizzen mast was also suspect and in need of attention. This was the smallest mast on board and it was decided that, rather than repair it, they would make a new one, so it too was taken ashore. A working party followed, who selected and fell a tree that would serve to make a new mast. When it was down, it was trimmed and taken to the place where the shipwrights were working on the foremast.

On the day the foremast was ready to be re-stepped, the wind rose to gale force which made it impossible to raise such a massive piece of timber with any degree of safety and it had to wait. The gale lasted for twenty-four hours but, once things had quietened down, the mast was successfully restored to its rightful position, complete with its maze of standing and running rigging. Once it was safely in place, it was the turn of the fore topmast to be raised, fixed and re-rigged. Meanwhile, the new mizzen mast had been made and, once it was stepped and rigged, that part of the refurbishment was complete.

Trade with the islanders continued and, once within reach of temptation, they proved themselves to be as skilful thieves as the Tahitians. As they possessed steel knives already, they were able to cut metal fittings from ropes. The recovery of stolen items proved to be more difficult than it had been in Polynesia. Cook, for once, seemed prepared to be lenient and to accept that a certain amount of petty thefts had to be tolerated. Charles Clerke, who was weakened by tuberculosis, was anxious not to waste his energies chasing thieves.

The naturalists had plenty to observe. While the refit of the vessels was taking place, they had the time and the opportunity to observe the rich plant and animal life. James Burney was just as fascinated by the attractive singing of the islanders, as by that of the birds. He attempted to note down both. William Bligh was busy sounding and surveying, while King and Bayly worked in their hastily set up observatory. Sometimes, Cook seemed to revert to his old self, going off in a boat with some of his midshipmen, rowing around the sound. One of these midshipmen, James Trevenen, noted one such occasion in his journal: 'Capt. Cooke… would sometimes relax from his almost constant severity of disposition, & condescend, now and then, to converse familiarly with us. But it was only for the time, as soon as we entered the ships, he became again the despot.'

All was ready for sea on 25th April and both vessels left their moorings on the ebb tide, at noon the following day, heading towards Alaska. Immediately, the wind fell away to nothing and the boats had to be put into the water to tow the ships out into the open sound. Slowly, the wind came back and *Resolution* and *Discovery* made their way into open water. Once there, the wind abruptly changed direction and increased in strength, a fairly sure sign of worse to come. And worse did indeed come. Heavy squalls, buried in hurricane force winds, plagued them for several days. Visibility was greatly reduced by the heavy rain and the overcast sky. Their first priority was to get away from the shore and its hazards. It took them some time to fight their way to deep water and safety. The bad weather went on for several days, forcing the ships to stay well offshore.

Being unable to sail close to the shore made it very difficult to carry out the original aim of the voyage, which was to find the elusive Northwest Passage. They were now a full year behind schedule. The coast of Alaska was an endless succession of inlets, most of which were very deep, and could have conceivably been the entrance to the famous passage.

As the foul weather began to abate, *Resolution* started to leak so badly that the pumps had to be manned constantly. It was not until 1st of May that the lookout sighted land again in latitude 55°20' north, about 500 miles to the north of Nootka Sound. Their need now was for sheltered water, so that they could deal with the leak in safety. Place names have changed and it is impossible to be absolutely certain where the repair was carried out, but it is likely to have been in the vicinity of Prince of Wales Island. They were unable to find a suitable beach on which to careen the vessel and they worked their way into shallow water and anchored. Cook then had the vessel heeled sufficiently to allow the shipwrights to reach the problem area. When the copper sheathing was removed they found the seams to be almost completely devoid of oakum. It is probable that *Resolution* had taken such a pounding in the heavy weather, that the oakum was torn out of the seams, a not uncommon occurrence for a wooden vessel, especially if the original work had been perfunctory. Once it was possible to reach the offending area, re-caulking the seams was not a difficult job and they soon had a watertight ship again.

The passage northwards continued, until the coastline began to turn to the west. Late in May a substantial inlet was found that headed off in a northeasterly direction. *Resolution* and *Discovery* sailed about 200 miles up this sound and then anchored. Some aboard *Resolution* were convinced that they had found the Northwest Passage, Cook was not and he chafed at the time lost exploring for it. From where they were anchored, boats were sent out to discover the extent of the inlet. Bligh was in charge of the boats and, on his return, he reported that the sound ended some miles further on, where it divided into two short arms, each of which terminated in a river. There was no way that it could be the Northwest Passage.

Cook extricated his vessels from the maze of islands at the mouth of the sound and then followed the coast to the south west. Although he was surveying at every opportunity, he failed to name the sound they had just left. When the expedition returned to England, Lord Sandwich named it Cook River, which has since been changed to Cook Inlet.

On 26th June they were sailing gently southwestward, in a light breeze and very poor visibility, when at about 1600 the sound of breakers was heard on the port bow. *Resolution* was immediately brought to and anchored in twenty-five fathoms. At the same time they called out to *Adventure* to follow suit. When the visibility improved, they found that they were within three-quarters of a mile of a small island, now called Unalaska, one of the Aleutian Islands which spread out in a curving line into the Bering Sea, pointing towards Siberia. The area surrounding the Aleutian Islands is notorious for its fogs.

William Anderson, the Surgeon, had contracted tuberculosis, probably before he joined *Resolution*, and had been getting progressively worse as the voyage wore on. Early in August, he finally succumbed to his illness. He and Cook had been good friends and Cook was distressed at his loss. He wanted to bury Anderson on one of the islands, but the time required to excavate a grave in that rock-bound land forced them to change their mind, and Anderson's body, sewn into a hammock and weighted with ballast, was buried at sea with due ceremony. Mr Law, the Surgeon from *Discovery*, was moved to *Resolution* and, Mr Samuel, the Surgeon's First Mate, was upgraded to Surgeon aboard *Discovery*.

On 9th August both barques were anchored in the shelter of Cape Prince of Wales, the westernmost point of America. Nine days later, the ships were almost 300 miles further north in latitude 70°44'. They were on the edge of the densely packed ice, which was ten to twelve feet high, and it was clear they were not going to make any more progress to the north.

There was no point in lingering in those conditions. It was obvious that their exploration was at an end for that season. Cook named the nearest point of land Icy Cape, and directed his course to the south, away from the wall of ice. He was unaware that when he turned away, he was only fifty miles southwest of the Beaufort Sea, which was the entrance to the Northwest Passage!

For some time Cook had been debating where they should spend the winter. He faced the usual problems: where could he get enough wood, water and fresh food to sustain the two hundred men for whom he was responsible? Not surprisingly he opted for the Hawaiian islands.

Before they could undertake the passage to Hawaii, it was necessary to deal with yet another leak, that had developed in *Resolution*. This time, it was the seams below the water line, well aft, on the starboard side, that were taking a great deal of water. When the shipwrights were able to remove the copper sheathing covering the leaky area, it was the same old story, many of the seams were wide open.

The ship's company were also in dire need of fresh food. Most of the vegetables they had picked up on their way north had been eaten or had rotted. On 2nd October they landed in the Russian trading port of Unalaska. The island was part of the Aleutian Group and the last point of land between Alaska and the Sandwich islands, almost three thousand miles to the south. Fortunately, there were a great many edible berries growing ashore that proved to be both tasty and a good anti-scorbutic. Spruce beer was brewed from young pine needles and this too kept scurvy at bay. The locals brought fish, both fresh and dried, to exchange for pieces of metal. Fishing parties were sent out from both ships so that in spite of the late season, the crews were well fed and stayed healthy.

Cook made contact with some Russians, who were trapping animals for fur, especially sea otters. These meetings were always agreeable, with an exchange of food to grace Cook's table and bottles of spirits of one kind or another given in return. The exchange of information was somewhat limited by the fact that neither side could understand the other's language but, as ever, food and drink smoothed the way, especially the drink. Before leaving Unalaska, Cook entrusted a departing Russian trader with a letter destined for the Admiralty in London. It was his first communication, since he had left the Cape of Good Hope, two years before. He describes the ice he encountered, tells of his intention to winter on Hawaii and then to return to the north in the following year to have another shot at finding the Northwest Passage. But, he sounds a pessimistic note, writing: 'But I have little hopes of succeeding.'

Resolution and *Discovery* sailed for Hawaii in the morning of 26th of October. As the wind came from the south, they were able to lay a westerly course. Two nights after they had left Unalaska, the main tack on *Discovery* parted and killed one man, injured the Boatswain and two or three others. The main tack is the rope restraining the bottom windward corner of the main course. The main course was probably the biggest sail in *Discovery* and the rope in question would have been carrying a considerable load. When a rope under tension breaks, its tail flies back like a released spring and woe betide anyone who is in its way.

At daybreak on 26th November, land was seen athwart their course. The island was Maui, although Cook was not aware of it at the time. It is obvious, though, that he knew he was close to the islands – he was steering for Kauai – because *Resolution* and *Discovery* had been working under reduced canvas throughout the night and, with dawn, he had the sail increased to enable them to close with the land they had seen.

With the wind astern, they were sailing directly for the island, and they altered their course to the west, which allowed them to sail parallel with the coast until noon,

when some canoes came out to them. The ships were hove to, to enable the canoes to come alongside. The occupants of the canoes clearly knew of Cook's earlier visit to Kauai and Niihau and boarded the two vessels without hesitation.

By the afternoon of 30th November, *Resolution* and her consort turned away from Maui to gain some sea room. When they had the offing that Cook wanted, the two barques were turned to the east, to sail parallel to the shoreline again, but in the opposite direction, and further away this time. Once past the northeast corner of Maui, they turned to the south in an effort to get to Hawaii and with the intention of sailing round that island. Unfortunately, the wind was right on their nose and it made it necessary for them to tack the length of Hawaii's northern coast. Each board took them only a very few miles in the right direction. It did, however, enable the canoes manned by locals and laden with fruit, vegetables and pigs to reach *Resolution* and *Discovery*. It was not until the end of December that the two ships were able to turn to the south and bring the wind abaft the beam, as they sailed along the south-east shore of Hawaii.

By 6th January 1779 they had weathered the southern-most point of the island and were sailing up its west coast. On the 7th, both vessels were anchored in Kealakekua Bay, nicely sheltered from the prevailing wind, with hundreds of Hawaiians swimming around them and many more in canoes. One report refers to the swimmers as if they were shoals of fish, there were so many. The officers reckoned that some 3000 canoes were assembled in the bay, with thousands of people packed in them, or swimming alongside. Cook was given the most tremendous welcome, it was almost worship. He stood on the quarterdeck, entranced by the homage, which must have made up for a lot of the frustrations of the voyage and for the debilitating effects of ill-health.

What he did not know was that the name of the bay – Kealakekua – signified 'pathway of the gods', and was considered to be the abode of Kanaloa, one of the four major deities of Hawaii. The bay was framed by cliffs, rising several hundred feet above the sea. These cliffs were the sacred spot, where the remains of dead chiefs were interred.

COOK'S SECOND VISIT TO THE HAWAII ISLANDS

Opposite: *Engraving of a white bear, based on a drawing by John Webber, who was to be employed by the Admiralty in compiling the official account of Cook's third voyage, published in 1784, and which includes this engraving.*

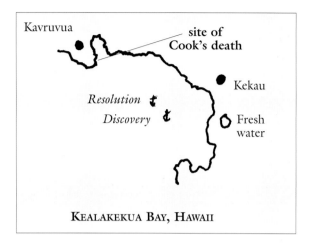

KEALAKEKUA BAY, HAWAII

It seems that the Hawaiians had convinced themselves that Cook was a god. By some extraordinary coincidence, Cook and his ships had entered Kealakekua Bay as the annual festival in honour of Lono, the god of plenty, was in full swing. During this festival, Makahiki, the symbols of Lono – white banners – had always been flown. *Resolution* and *Discovery* sailing into the bay with billowing sails must have looked truly like Lono's vehicle. The priests seemed unsure of Cook's status, but the Hawaiians seemed determined to receive the god in their midst. They climbed on board, in such numbers that they practically swamped the vessels. Anything, that was not bolted to the decks was stolen, so eager were the locals to possess a relic of the god's visit.

This adulation seems to have had a strange effect on Cook. The day they sailed into Kealakekua Bay, 17th January, marks his final entry in his journal. This was going against the habits of a lifetime. From then on we have to rely on Lieutenant King and Lieutenant Phillips of the Royal Marine, for an account of events in Kealakekua Bay.

The ships' companies were surprised, when they came on deck on the morning of 24th January and found that there were no Hawaiians to be seen. Later that day, they learned that the bay had been declared 'tapu' (which has been anglicised as 'taboo'), on the occasion of the imminent arrival of Kalniopu, Chief of all the Hawaiian islands. He soon arrived on a private visit, accompanied by members of his family. He first boarded *Discovery* and then moved on to *Resolution*.

Whenever Cook stepped ashore, one of Lono's priests escorted him through the throngs of people murmuring the name of Lono. The chiefs, even the old and revered Kalniopu, treated him as a god and seemed to think him immortal.

The next day there was a full ceremonial visit by the same chief. Lavish presents were given to Cook, the most important being a cloak and helmet made of the red feathers that were held in such high esteem throughout Polynesia. Cook responded

Opposite: *James Cook arrived on Hawaii during the festival of the God Lono. This fact convinced the islanders that Cook himself was an incarnation of the God. In this picture a pig is being sacrificed to Cook, who is seated on the right of the standing priest. Behind them are the hut and wooden images of the* hikiau *or temple.*

in kind, even unbuckling his sword and belt and placing them around the chief's waist. Cook and the chief exchanged names which, throughout the Pacific, was considered to be the highest mark of friendship that could be offered. It was not until the ceremonies were at an end that the islanders reappeared. For some three weeks, Cook was feasted and honoured at every turn.

But gradually, the mood began to change. The sailors, with their relentless pursuit of local women and their drunkenness, were beginning to look like strange company for a divine being. Furthermore, an elderly sailor, who had followed Cook from retirement at Greenwich Hospital, died suddenly. This proved to the Hawaiians that the strangers were by no means immortal. Lieutenant King's worry that things could turn sour, 'should this respect wear away from familiarity, or by length of intercourse,' was proving to be well-founded.

The Hawaiians treated the ships' companies with great kindness but the continued burden of feeding them was beginning to weigh heavily on those who had to provide the food. The celebration of Makahiki had always included great offerings of food, but never on this scale. No matter how many supplies were brought to the

ships, more were always required. Cook let it be known that he would be leaving before long, and in response, Kalniopu, ordered his people to provide a vast amount of food as a parting gift for the two ships. This was done and the islanders must have breathed a sigh of relief, when the vessels finally raised their anchor.

It was early in February, when they left Kealakekua Bay. It was Cook's intention to complete the survey of the islands and then make his way to the Bering Straits. The survey of the Hawaiian islands was a slow business, chiefly because the winds were so light.

On 7th February the weather changed dramatically to violent gale-force winds, with even stronger squalls, lasting for thirty-six hours. In the course of the storm *Resolution*'s foremast developed a shake, which is a long vertical crack. The damage was such that it was essential to take the mast out of the ship and put it ashore to enable the shipwrights to repair it. Not surprisingly, Cook was extremely angry, had the dockyards in England done their work properly, the recurring problems with the ships would not have arisen. This was nothing new, but Cook was taking the latest difficulty very badly indeed. It was obvious to those who knew him well that he was tending to over-react to situations he would have taken in his stride on a previous voyage.

Cook had to decide where he should go to carry out the repairs, and be safe from the weather, should it deteriorate again. The foremast was nearly as long as *Resolution* herself. Removing a spar of that size and placing it ashore, without dockyard facilities, would be a major exercise in seamanship. It was essential to carry out the work in a sheltered spot. His survey of the islands was incomplete, he had intended to finish it as he sailed away. The only place he could be sure would offer him the shelter and the timber he required, was the one he had just left. Three days later, the ships were back to a muted reception and anchored in the same spots as before in Kealakekua Bay.

The astronomers re-erected their tent ashore in its previous position, with the apparent approval and friendly hospitality of the priests. Fresh food was brought to the ships, as freely as before, but there was an underlying sense of unease. Kalniopu visited Cook. He was obviously annoyed that the vessels had returned and Cook explained the matter of the broken mast. The chief was irritated, but willing to allow some of his men to help with the repairs.

It took a further three days to unstep the mast and get it ashore, where they found that at least eight feet of its lower end were rotted. On 13th February work on the mast was starting. A little way off, there was a party of seamen collecting water and, as before, they had engaged some of the local people to help. Suddenly, a number of local chiefs ordered the locals, who were working with the watering party, to withdraw. It was as if they had waited for the mast to be taken out of *Resolution*, which rendered her helpless, before driving the Hawaiians away. A man stole a pair of tongs, was caught and Clerke had the culprit given forty lashes.

Not long after, the chiefs and some other men became increasingly offensive and eventually started to throw stones at the working party. An officer, who happened to be at the observatory, walked down to the scene of the disturbance with an armed marine and this was enough to restore order and allow the men to resume their work.

The minor fracas attracted Cook's attention and he made his way ashore. As he landed, shots were fired at a canoe that was fleeing from *Discovery*. Cook ran along the shoreline in an attempt to catch the occupant of the canoe, but was too late. The thief had stolen another pair of blacksmith's tongs and a chisel. Cook, King and a marine set off in pursuit, but failed to catch him. No one ashore knew it at the time, but, in fact, the stolen tools had already been recovered.

In Cook's absence, Edgar, the Master of *Discovery*, seized the canoe the thief had used. Unfortunately the canoe belonged to a chief who lived close by. This chief had been a particular friend of Cook's and had gone out of his way to be helpful to both the English ships. Understandably, he objected strongly to his canoe being taken in this arbitrary fashion.

By now, a sizeable crowd had gathered and the chief grabbed Edgar's arm to remonstrate with him. A seaman who had been with Edgar, knocked the chief down with an oar. The chief, a powerful man, took the oar from the sailor and broke it in two. His action spurred the crowd on to attack the offenders. In no time at all, there were stones flying and the seamen, who were unable to launch the pinnace, were glad to swim to a rock that put them beyond the range of the stone throwers. Two of the seamen who could not swim were protected by the angry chief, who pacified the stone throwers, helped the seamen to launch the pinnace and gathered together enough oars for the crew to make their way back to *Discovery*.

Captain Cook followed later, and it was not until he was aboard, that he was able to piece together the full story. He was furious over every aspect of the whole sorry affair. He was angry with Edgar for taking precipitate action, he was angry with the thief, he was angry with himself for not catching the thief, and he was angry with the chief and his people. He was so angry that he threatened to take reprisals that he was not really in a position to carry out due to the inequal numbers between his crew and the islanders, a mistake he would not have made before.

Cook clearly had not come to terms with the idea that the Hawaiians were not so easily cowed as the Tahitians. It is possible that part of the problem was that the Hawaiians were not fully aware of the power of the fire arms carried by the marines. Cook told King that 'these people will oblige me to use some violent measures. For they must not be left to imagine that they have gained an advantage over us.' So far, the only shots that had been fired in Kealakekua Bay had been aimed at the fleeing canoe and they had missed. Yet, the sheer force of number far outweighed anything the marines could have done.

It was the practice that whenever boats needed to be left afloat overnight, they would be filled with seawater to frustrate potential thieves. At dawn, on 14th February 1779, they noticed that *Discovery*'s cutter, quite a large boat, was missing. When Cook was informed, he decided to take Kalniopu on board *Resolution* to encourage the thieves to return the cutter, a tactic that had worked often enough in Tahiti. When the chief was found, he reluctantly agreed to go with Cook, but one of his wives and two lesser chiefs strongly objected and physically restrained him. By now, a huge crowd had gathered, one report says 3,000 people. It would have been the time for Cook to swallow his pride and return to *Resolution*, with what dignity he could muster and without his hostages, but it was not in his nature to accept defeat.

The party of eleven marines was ashore, in a line close to the water's edge and, behind them, some twenty feet off, were the launch in the charge of Lieutenant Williamson and Cook's pinnace in the charge of Roberts, the Master's Mate of *Resolution*. This was a poor arrangement, as the marines found themselves directly in the line of fire of the men in the two boats. Cook was perhaps twenty-five yards ahead of the line of marines, alone, still talking with Kalniopu.

It was becoming increasingly apparent that the Hawaiians were armed and hostile. Cook realised that, if he attempted to take the chief aboard, there would be considerable bloodshed, something that he wanted to avoid.

Whilst this was going on, a canoe that was trying to leave the bay was forced ashore, shots were fired and an important chief was killed. The news spread quickly among the vast crowd and it did nothing to pacify the Hawaiians. One man, bolder than the rest, threatened Cook with a pahua. Ironically these weapons had been made on board *Resolution* and *Discovery*, as items of barter. The pahua was a tapered iron bar an inch in diameter and generally about two feet long, ending in a sharp point. Cook discharged his musket, which was loaded with small shot, at the man, but to no effect, the protective mat he was wearing, saving him from harm.

A chief also armed with a pahua, attacked Lieutenant Molesworth Phillips, the officer in charge of the marines. Phillips struck the chief a disabling blow with the butt of his musket. The mob surged forward, throwing stones and brandishing spears and clubs. Cook fired and killed a man. Phillips was attacked again. First he was felled by a stone and then stabbed with a pahua, but he managed to shoot his assailant before the man could strike again.

Cook waved to the boats and ordered the men to get into them. The crew in the boats appear to have been confused by Cook's wave: the pinnace attempted to come closer to the shore, but Williamson ordered the launch to back off into deeper water. One would have thought that he would have used his initiative and returned to the water's edge to attempt to pick up the men who were stranded there and under attack, but he may have thought that Cook had waved him off and, discipline being what it is, he obeyed the 'order'.

While Cook faced the mob, no one attempted to molest him but, once he turned and tried to reach the boats, he was struck on the head with a club and went down. As he lay on the ground, partly in the water, he was attacked by numerous Hawaiians, almost as if they wanted a share in the glory of killing him. The guns on *Resolution* were brought to bear and a number of rounds were fired which, with the musket fire, cleared the beach. But it was all too late: Cook and four marines lay dead. At this point, the boats could have gone in to the beach and recovered the bodies. The seamen and the marines were anxious to do this, but the officer in charge, Lieutenant Williamson, ordered the boats back to *Resolution*.

Throughout the incident, Clerke had been powerless to go to the aid of his friend. He had no boats and one was needed in order to attach a spring to the anchor cable, to let him slew *Discovery* and bring her guns to bear on the islanders. (The carriage guns would normally be set up to fire only at right angles to the fore and aft line of the ship.) He was also unable to send an armed party ashore. Had he succeeded in either of these two actions, it probably would not have influenced the outcome, but caused a few more deaths on either side, all to no purpose.

The Hawaiians took Cook's body, dismembered it and distributed the various parts among their chiefs. Much of the flesh was burned and the bones were kept as trophies.

Responsibility for the two ships and the welfare of their crews now devolved upon

Captain Clerke. He shifted his berth to *Resolution* and appointed Lieutenant Gore to command *Discovery*. The tuberculosis Clerke had picked up in gaol was causing him a great deal of suffering and he was a very sick man indeed. Yet, the security of the two barques, and of everything they had ashore, were Clerke's immediate responsibility. Most of the sails from both vessels were still ashore for repair and airing, so was the astronomers' scientific apparatus, all of which had to be safely retrieved. The marine guard had a large tent which had to be dismantled and returned aboard and, of course, the unfinished foremast had to be completed and reinstalled.

Each day, casks had to be filled with water and the men doing this or collecting the gear, were subjected to stone-throwing and insults. The marines did what they could and, occasionally, the ships' guns were brought into action, but the harassment continued. Each night after dark, guard boats were rowed around the two ships to deter attack from the shore.

Eventually, the men working ashore could no longer tolerate the locals' provocations and they asked Clerke for permission to retaliate. Clerke received their request with sympathy, but was reluctant to give his formal permission. Since action had not been vetoed, the men took it upon themselves to believe they had been given permission to act. The next day, when the harassment and the stone-throwing started again, the men turned on the Hawaiians and drove them off, inflicting considerable losses. For good measure, they burned a number of nearby houses and groves of fruit trees.

Sad as it may seem, it did the trick. From then on, the working parties were left in peace. Kalniopu sent an envoy to seek peace and canoes began to arrive with fresh food to barter.

Later on, a solemn procession of chiefs, bearing presents of food, approached the beach and, with them, was Kalniopu's envoy, carrying a large package wrapped in the finest cloth and the whole covered with a cloak of black and white feathers.

The package contained the remains of Captain Cook. The hands were included and this made formal identification simple, as one of Cook's hands had been mutilated by an exploding powder horn, many years before, in Newfoundland. The chiefs were told by Clerke that he wanted the bay to be made strictly 'tapu' for the following day and they agreed readily.

Opposite: *The death of Captain Cook in Hawaii, on 14th February 1779*

On 22nd February 1779, before the mournful backdrop of the fire-blackened ruins of the houses and trees that had been burned the day before, the lower deck was cleared. With an occasional wisp of smoke still rising from the ashes in the calm of the evening, all the flags aboard the two ships were put at half mast.

On *Resolution*'s upper deck the officers, arrayed in their best finery of blue, white and gold, the Royal Marines dressed in their scarlet jackets and white breeches and the ranks of the seamen in their less colourful gear, were drawn up in silence. It was a sombre gathering, many of the men had been with Cook on all three voyages and held him in high regard. One can well believe that there was many a tear shed for him that day.

While the minute-guns thundered out their salute, amidst billowing clouds of pungent smoke, Captain James Cook, a brave and brilliant man, a much-loved seaman and revered scientist, was committed to the sea with all the pomp and honour that could be given him.

On 22nd February 1779 the two vessels finally sailed out of Kealakekua Bay. Clerke was so ill that he had to depend on William Bligh for the navigation. Whatever their grief and shock, they had their orders to carry out and they were sailing back towards the icy wastes, still intent on finding the Northwest Passage, but with little hope of achieving this aim. The heart had gone out of their exploration.

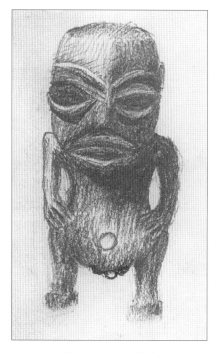

Polynesian idol

Overleaf: *General chart prepared by Lieutenant Henry Roberts of the Royal Navy, showing Captain Cook's track during his three successive voyages.*

22. Epilogue

Eleven months after Cook's death, the *London Gazette* published the following account in its issue dated 11th January 1780: 'Captain Clerke of His Majesty's Sloop Resolution, in a letter to Mr Stephens, dated 8th June 1779... which was received yesterday, gives the melancholy account of the celebrated Captain Cook, late commander of that Sloop with four of his private Marines having been killed on 14th February last at the island of O'Why'he, one of a group of new discovered Islands in the 22nd Degree of North Latitude, in an affray with a numerous and tumultuous Body of Natives.'

By the time Clerke's letter reached Mr Stephens at the Admiralty, Clerke himself had succumbed to consumption and been buried in Kamchatka, the Russian peninsula, situated between the Sea of Okhotsk and the Bering Sea. All of Cook's friends: Lord Sandwich, Sir Hugh Palliser, Sir Joseph Banks, and so many others, were devastated by the news. Elizabeth Cook, who had lived all her married life in the fear that her husband might not make it back, was now a thirty-eight-year-old widow. It is not known how she found out about his death. Was a messenger sent by the Admiralty, did one of Cook's supporters go to Mile End in person to break the news? Three-and-a-half years had elapsed since she had last seen her husband. Their last child, Hugh, was now a small boy. Their two other surviving sons, James and Nathaniel, were in the Navy, having completed their training at the Portsmouth Academy.

George III was greatly moved on learning of Captain Cook's demise and granted a pension of £300 per year to Elizabeth. One of Cook's recurrent worries had been his wife's financial situation, should something untoward happen to him. Posthumous honours, pensions, the income from the publication of Cook's journals ensured that his wish had been granted: Elizabeth would want for nothing, at least, materially speaking.

Dressed in mourning till the day she died, Elizabeth would survive her husband by fifty-six years. In the course of her long life, she was to endure many more tragedies: Nathaniel went down with the ship aboard which he served as a Midshipman – the *Thunderer* – during a hurricane off the coast of Jamaica in 1780, about a year after his father's death. The youngest son, Hugh, who was to enter Holy Orders, died of a fever while still studying at Christ's College, Cambridge, in December 1793. James, the eldest, who had been recently promoted to Commander of the *Spitfire*, died one month later, apparently the victim of a robbery. Elizabeth went to live with a cousin, Isaac Smith, who was a retired Rear-Admiral living in South London. He

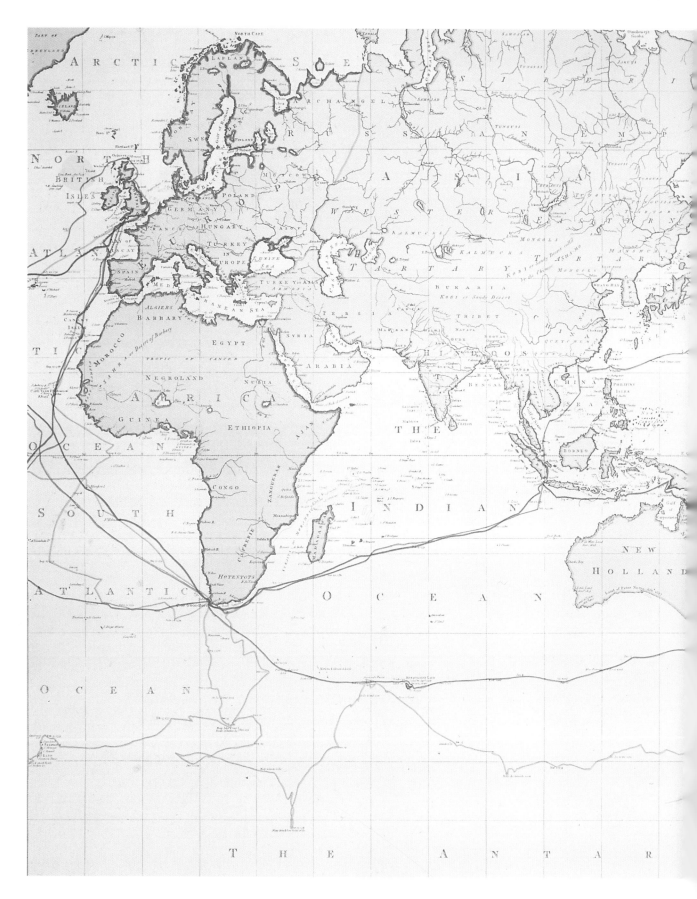

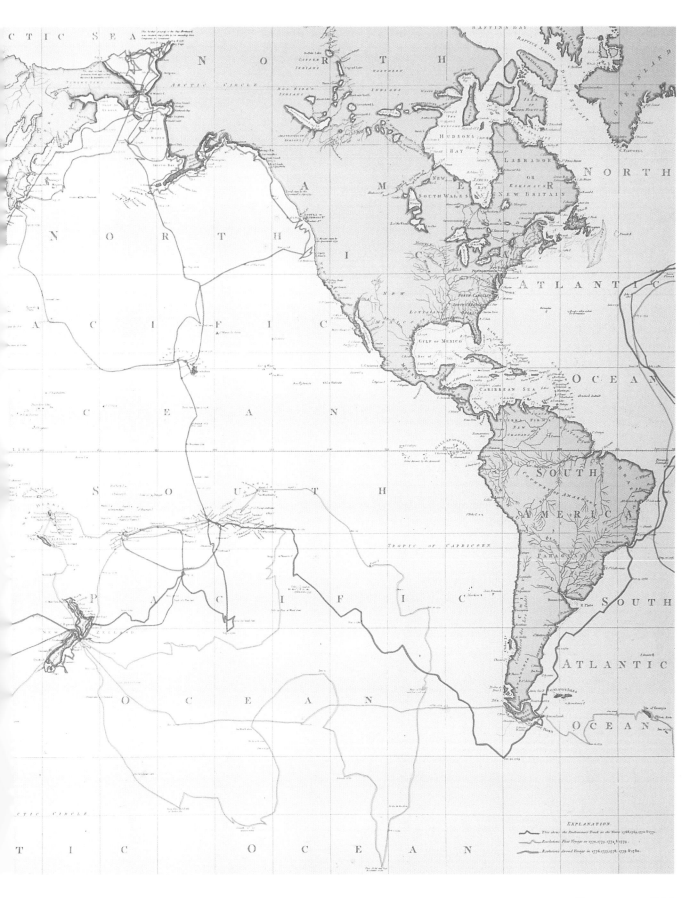

died eventually and Elizabeth went back to her own house. As mentioned before, she burned all of her husband's and her personal letters to him, shortly before she died at the age of ninety-three. She was interred under the floor of the Church of St Andrew the Great, in Cambridge, along with two of her sons: Hugh and James. In over sixteen years of marriage, she and Cook had spent exactly four years together.

In his lifetime, Cook had achieved an amazing popularity and influence. This was all the more remarkable in view of his humble beginnings in an age not known for upward social mobility. Luckily for him, the Navy and the Admiralty often did recognize outstanding talent when they saw it. His name is still widely respected throughout the world today. There have been various attempts to topple him off his pedestal, on the basis of diminishing faculty and acts of cruelty committed during the last fatal voyage. Disturbing and unpleasant events undoubtedly took place, some of which probably contributed to his ultimate violent death. A variety of latter-day diagnoses have been put forward by various authorities to explain the dramatic personality change undergone by Captain Cook. Although based on the detailed observations noted in the journals of some of his companions on the last voyage, these theories remain speculative. Whatever the reasons, Cook and his behaviour changed radically and the idol revealed that it was but human after all, and to state this is not to denigrate his very real achievements. Despite the errors of the last voyage, I still regard him as an outstanding seaman and a genuine humanitarian.

Among his many achievements, and possibly the one he himself ranked most highly, was his success in eradicating the ravages of scurvy, and, speaking as a limey, I am proud to say amen to that. (Incidentally, the word 'limey' is an interesting nineteenth-century American and Canadian slang describing a British sailor, ie one who drinks lemon juice as a protection against scurvy.) Of course, the knowledge that enabled Cook to bring about this change had existed, long before he was in a position to influence matters, but it had never been put into systematic practice. The earliest reference I have found to a possible cure for scurvy dates back to 1593 and Sir Richard Hawkins. He used citrus juice against scurvy amongst his ship's company and it is possible that others could have used the same method at an even earlier date.

Captain James Lancaster unintentionally conducted a controlled experiment in the use of citrus juice, as a specific against scurvy, aboard his flagship *Dragon* in 1605. He was making a voyage to the East Indies in company with three smaller vessels. Lancaster used citrus juice to keep his ship free of scurvy, but the other three ships did not. They lost 105 out of a complement of 222 men, all due to scurvy. James Woodall, in his book *The Surgeon's Mate*, published as early as 1617, states that citrus juice included in seamen's diet would prevent scurvy.

What was needed was a humanitarian who was sufficiently disturbed by the sickness and death caused by this acute vitamin C deficiency, to want to do something about it: one who also had an enquiring mind and sufficient power of command, laced with a measure of guile to enable him to get his ship's company to accept the measures necessary for their well-being. Before Cook came along, the Navy seemed to regard scurvy as an inevitable part of life at sea and one which was expected to kill more men than wars and rough seas ever did.

Until 1850 the Hawaiians held Cook's memory in high esteem, perhaps something of their belief in him as a quasi-divine figure had subsisted. All this was to change with the advent of the missionaries who arrived on Hawaii in that year, and those who followed. They saw Cook as an obstacle to their intention to convert the Hawaiians to Christianity. The missionaries, a Reverend Sheldon Dibble, in particular, set about the denigration and destruction of Cook's character to further their own ends. It seems that Dibble had no qualms about lying and wilfully distorting historical facts to serve his own cause. He converted the Hawaiians and trained them as teachers and pastors, while indoctrinating them with his version of history. They in turn misled their flocks in the same way, until Cook became an object of condemnation amongst Hawaiians. It is sad that such a great man should have suffered posthumous character assassination by one who should have had a greater respect for the truth.

There are memorial stones and tablets scattered all round the Pacific in praise of Cook. Even as far afield as Anchorage, Alaska, one can find a statue to his memory, but, until the early part of the twentieth century, the only memorials in Hawaii were erected by visiting Royal Navy ships' companies. The centenary of Cook's discovery of the islands was marked by the erection of a statue of chief Kameha'meha who, it is believed, played a leading part in the murder of Captain Cook.

By 1928 the malign influence of the early missionaries had largely dissipated and the sesquicentennial celebration of Cook's arrival in the Islands of Hawaii was celebrated in grand style. The Hawaiians were joined for the ceremonies by dignitaries from Australia, New Zealand, the United States and Britain. Pageants were staged, a memorial was unveiled at Kealakekua and another at Waimea, and commemorative coins were struck to mark the occasion.

Today, Cook is accepted by the islanders as part of their history and this is as it should be. It is good to be able to end on this happy note.

Glossary

Aback: Wind in the wrong side of a sail. It may be unintentional, or used to drive a vessel astern in tight quarters. This manoeuvre is also used to cause an anchor to dig into the seabed, after it has been dropped: the vessel moves backwards to lay out the cable and the anchor digs in.

Abaft: Immediately behind

Abeam: Something ashore or afloat that is at ninety degrees to the **fore and aft line** of a vessel, eg a wind is said to be **abeam**, when it blows square onto the side of a vessel. Another ship or an object ashore or afloat is said to be **abeam**, when it is to one side of a vessel.

Anti-scorbutic: Vitamin C-rich food or drink that helps prevent or cure scurvy.

Athwartship: At right angles to the **fore and aft line**

Atoll: A reef and/or small islands encircling a lagoon and built by coral polyps and associated organisms.

Beam: The dimension of a vessel at its widest part

Bend: Some knots have the word **bend** as part of their name, eg 'sheet bend,' 'carrick bend' etc. To **bend** something is to tie it to something else, eg a sail is **bent** to a spar.

Bent: Past tense of **bend**

Bilge: The lower part of a vessel, both inside and out

Blocks: See **sheaves**.

Opposite: *Memorials to James Cook are to be found at various sites in the territories he discovered. This particular one was photographed by a relative of the author, David Jones, in the town of Anchorage, Alaska. The statue is the work of Derek Freeborn, who based it on the memorial to James Cook in Whitby, where Captain Cook started his sea career.*

Boards: When sailing the zigzag course necessary to take a vessel to windward, each leg is called a **board**. Most square-rigged sailing vessels could sail no closer than 60° to 70° degrees to the wind.

Boom: A spar used to maintain the shape of the foot of a sail.

Bower anchor: A vessel's main anchor

Cable: A rope used with an anchor is a cable. A **cable** also means a tenth of a nautical mile, or 200 yards.

Careen: To cause a vessel to heel over, when sailing, so that the underwater parts are exposed. It also describes the act of causing a vessel to heel until its underwater section is exposed to allow work to be carried out on that area, without the need to haul the vessel out of the water. Normally done by shifting weights in the vessel and/or hauling on a line, attached to the masthead.

Catamaran: A double-hulled vessel

Caulking: Driving cotton or oakum between the planks on the hull and deck of a vessel to make it watertight.

Centre plate: A wooden board or metal plate that could be raised or lowered within a water-tight box. When lowered, it would give the boat enough lateral resistance to enable it to be sailed. Raised, it would allow the boat to float in very shallow water, but having lost its centre plate, it could only sail downwind.

Cranky: A vessel that heels unduly, maybe dangerously, to the least wind and/or does not respond to its rudder as it should.

Cutter: A sailing vessel with a single mast and two sails forward of the mast. Before the introduction of small engines, the Navy used **cutters** as general work boats. They were open boats, with a centre plate, about twenty-eight feet in overall length. They could be rowed or sailed.

Deadwood: The timber that spans the space between the underside of the hull and the keel. So called because there is no space within it.

Draught: The greatest depth of the underwater section of a vessel

Fathom: From the old Scandinavian 'fadom' meaning 'embrace.' When the leadsman hauled his lead line aboard, he would measure its length by stretching each length as it came aboard, to the full extent of his arms. The distance between the hands, when the arms are extended on either side, is approximately six feet. The action was reminiscent of embracing someone repeatedly.

Fore and aft line: An imaginary line that runs from the stem head to the middle of a vessel's stern.

Foremast: The foremost mast in a vessel

Fringing reef: An area of reef, usually coral, that lies detached from, but parallel with a shoreline.

Good holding: A section of the sea bed that will allow an anchor to dig in and hold the vessel securely.

Headed: When the wind direction alters, making it impossible to hold the course, the wind is said to have **headed** the vessel.'

Heave to: A gambit that will stop a vessel and keep her roughly in the same position. Aboard a sailing vessel it is achieved, usually by backing some of the sails, whilst others are left drawing, this results in the vessel being driven forward by those sails left drawing, and driven backwards by the sails that have been backed, resulting in a stationary vessel.

Holystone: to scrub a vessel's decks with a **holystone**, which is a soft sandstone.

Kedge anchor: An anchor which is smaller than the vessel's main anchor. Often used to **kedge off.**

Kedge off: To carry an anchor out from a vessel, in the direction in which it is intended to move that vessel, by hauling on the **kedge anchor** cable, once that anchor is dug in.

Larboard: The term in use in Cook's day for the present day **port**.

Lateen sail: A triangular sail supported by a spar on its leading edge. Sometimes a spar is also **bent** to the lower edge of a sail, as in the traditional Polynesian sail.

Lay to: In heavy weather, it may be necessary to lower most of the sails, retaining just a fore and aft sail, set in the after end of the vessel, to keep her head up towards the wind. This enables the vessel to meet the oncoming seas in the best possible way.

Lee shore: A shore onto which the wind is blowing. A dangerous position for a vessel, because it is so difficult to sail away from.

Leeward: The lee-side of a vessel is the side opposite to that upon which the wind blows.

Main mast: The tallest mast in a vessel

Mizzen mast: The aftermost mast in a vessel

Missed stays: If, when a sailing vessel is changing from one tack to the other, it fails to cross the wind to the new **tack**, it is said to have **missed stays**.

Mizzen staysail: A staysail is a fore and aft sail. Sails take part of their name from the mast upon which they are rigged. A stay is a wire, or rope, that is secured to a mast to give fore and aft support, in this case the **mizzen mast**. A **staysail** would be rigged on this wire, hence **mizzen staysail**.

Neap tide: Tides have a fourteen-day cycle, during which they rise to greater and lesser heights. The highest tide in the cycle, the spring tide, follows on the full moon and the new moon. The height of the tide falls each successive day for seven days, when it reaches its lowest level, this is the **neap tide**. The low-water level is also affected. In a **spring tide** the low water will be the lowest in the fortnight. In a **neap tide** the low water will be the highest in the cycle.

Off and on: The act of sailing towards the land and then away from it, repeatedly. It is done usually when a sailing vessel is approaching land in darkness and does not wish to close with the coast until daylight.

Overfalls: See **rip tide**.

Pay off: To allow the vessel to fall away from the wind.

Pay out: To allow a rope to run out under control.

Point: An old compass reference equal to eleven and a quarter degrees.

The early compass card was divided by a north/south line and an east/west line, making four ninety-degree quadrants. These were further subdivided until there were thirty-two divisions, each measuring eleven and a quarter degrees. Also used to indicate that the wind allows the vessel to '**point** up into the wind,' far enough to be able to lay her course. A further use was the order to '**point** her up a bit,' meaning to sail closer to the wind.

Port: The left-hand side of a vessel, when facing forward

Quarter: The two after 'corners' of the vessel, less often used to refer to the **port** or **starboard** bow sections.

Rip tide: Also **tide race** and **overfalls**. A stretch of the sea where the sea bed is uneven. This causes the surface of the sea to have large and irregular waves, that can be dangerous, especially to smaller craft.

Running rigging: Rigging that moves, eg a rope used to raise or lower a sail.

Running survey: A survey carried out from the ship, as it sails along the coast

Scarph: To join lengths of timber to make one long piece, such as a mast. This may be a simple slope at the ends of the timbers to be joined, or it may be a complex interlocking joint, if greater strength is needed.

Schooner: A sailing vessel with two masts, the taller – the main mast – is placed aft. The sails are normally fore and aft sails, but a jackass **schooner** also has one or two square sails on the fore top mast.

Serve (wind): A wind that blows so that the vessel can sail the chosen course, is said to **serve**.

Shallop: An open boat designed to operate in shallow water

Sheave: What is called a pulley ashore, is known as a **block** at sea and the grooved wheel contained within the body of the block, is the **sheave**.

Slew: A rope attached to the anchor cable and brought to the side of the vessel will, when pulled, cause the vessel to turn – to **slew** – to one side.

Spar: A pole used for one of a number of purposes on board ship. In a sailing vessel it is often used to help spread a sail, so that it can be filled by the wind. The masts and yards are also called **spars**.

Spring: A rope rigged at an angle to the fore and aft line. When moored alongside, a **spring** is rigged to stop the vessel surging forwards or backwards. Hauling on the **spring** is a method which is sometimes used to help manoeuvre a vessel. A **spring** tied low down to the anchor cable and brought aboard at about midship, or further aft, can be used to **slew** the boat round, if the anchor cable is slackened off.

Spring tide: See **neap tide**.

Standing rigging: Rigging that does not move. See **running rigging**.

Standing on and off: When approaching a landfall in the dark and which has no illuminated buoys, a sailing vessel would be sailed back and fore, a safe distance from the landfall, until sunrise. See **off and on**.

Starboard: The right-hand side of a vessel when looking forward. Early vessels had the rudder hung on what is now the **starboard** side of the stern, the rudder was called the 'steerboard.' That side of the vessel became the 'steerboard side' and, eventually, **starboard**.

Stock: The rudder **stock** is a piece of timber that forms the leading edge of the rudder blade and extends upwards into the ship, where it is fitted with a tiller, or tiller lines, that connect it to the steering wheel. It is also the horizontal bar in the traditional anchor.

Studding sails: Additional lightweight sails that are set in light airs.

Tack: If the sails are set so that the wind fills them from the port side, the vessel is said to be on the **port tack**. The opposite would be the **starboard tack**. To change from one **tack** to the other is to steer the vessel in such a way that her bow crosses the wind.

Tide race: See **rip tide**.

Tiller: A 'handle' fixed to the rudder head to enable the helmsman to turn the rudder.

Top gallant mast: In the barque rig with which Cook's vessels were equipped, the top gallant mast would be the top most mast on each of the masts.

Warp: A rope used to moor, anchor, tow or manoeuvre a vessel.

To warp: To manoeuvre a vessel by hauling on the ropes.

To wear ship: To put a vessel on to the other **tack** by bringing her round, stern to the wind. It achieves the same purpose as tacking but can lose more ground than tacking. The weather can be too bad for a vessel to be **tacked**, but **wearing ship**, will always succeed.

Windage: The whole of that part of a vessel that is above the sea level.

Wort: An anti-scorbutic drink made from malt, steeped in boiling water.

Yawl: A two-masted sailing vessel in which the **mizzen mast** is stepped **abaft** the tiller.

Palm trees in the trade winds

Bibliography

Mr Bligh's Bad Language: Passion, Power and Theatre on the Bounty, Greg Denning, Cambridge University Press, 1992

Captain Cook: The Life, Death and Legacy of History's Greatest Explorer, Vanessa Collingridge, Ebury Press, London, 2002

Captain Cook, the Seaman's Seaman, Alan Villiers, Hodder and Staughton, London, 1967

Captain Cook's Final Voyage – The Journal of Midshipman George Gilbert, Christine Holmes (ed), Caliban Books, 1982

Captain James Cook – a Biography, Coronet, 1995

Captain James Cook and his Times, Robin Fisher & Hugh Johnston, Douglas & Macintyre/Croom Helm, London, 1979

Cruising the Coral Coast, Alan Lucas, Castle Books and Gordon & Gotch, 1982

A Dictionary of Sea Terms, A Ansted, Brown, Son and Ferguson, Glasgow, 1933

English Social History, G M Trevelyan, Longmans

The Explorations of Captain James Cook in the Pacific, A Grenfell Price (ed), Angus & Robertson Ltd, London, 1969

Farther Than Any Man, Martin Dugard, Washington Square Press, part of Simon and Schuster, 2001

The Haven Finding Art, E R G Taylor, Hollis and Carter, London, 1956

His Majesty's Bark Endeavour: The story of the ship and her people, Antonia Macarthur, Angus & Robertson in association with the Australian National Maritime Museum, Sydney, 1997

A History and Practice of Navigation, R D Hewson, Brown, Son and Ferguson, 1963

Into the Blue, Tony Horwitz, Bloomsbury Publishing, London, 2002

Joseph Banks, Patrick O'Brian, Harvill Press, London 1997

Life of Captain Cook, Hugh Carrington, Sidgwick and Jackson Ltd

The Life of Captain James Cook, J C Beaglehole, A & C Black, London, 1974

Longitude, Dava Sobel, Fourth Estate Ltd, London, 1996

The Murder of Captain *James Cook,* Richard Hough (MacMillan) 1979

Maritime Meteorology, G E Earl and N Peter, The Maritime Press, 1968

Nature's Government: Science, Imperial Britain and the Improvement of the World, Richard Drayton, Yale University Press, New Haven and London, 2002

Preserving the Self in the South Seas, 1680 to 1840, Jonathan Lamb, University of Chicago Press, 2001

Red Sea and Indian Ocean – Cruising Guide, Alan Lucas, Imray, Laurie, Norie and Wilson Ltd, St Ives, Huntingdon, 1985

The Ship – Retracing Cook's Endeavour Voyage, Simon Baker, BBC Publications, 2002

Shoal of Time, Gavan Daws, University of Hawaii Press, Honolulu, 1974

Terra Australis to Australia, Glyndwr Williams & Alan Frost (ed), Oxford University Press, in association with the Australian Academy of the Humanities, Melbourne, 1988

Two Worlds: First Meetings between Maori and Europeans, 1642-1772, Anne Salmond, Viking, Auckland and London, 1991

Voyages of Discovery, Captain Cook and the Exploration of the Pacific, Lynn Withey, Hutchinson, London, 1967

The Voyage of Endeavour 1768-1771, Vol 1, J C Beaglehole (ed) for the Haklluyt Society, Cambridge University Press, 1968

The Voyage of Resolution and Adventure, Vol 2, J C Beaglehole (ed) for the Haklluyt Society, Cambridge University Press

The Voyage of Resolution and Discovery 1776-1780, Vol. 3, J C Beaglehole (ed) for the Haklluyt Society, Cambridge University Press

The Voyage of the Endeavour – Captain Cook and the Discovery of the Pacific, Allen &Unwin, 1998

Voyages and Discoveries, Richard Hakluyt, William Collins & Sons, London, 1975

We the Navigators, Dr David Lewis, Australian National University Press, 1972

The Wooden World: An Anatomy of the Georgian Navy, N A M Rodger, William Collins & Sons, London, 1986

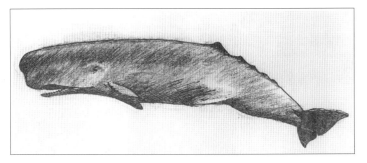

Whale

Index

Figures in Italic refer to illustrations.

Aborigines: 97
Admiralty: 24, 26, 29, 51, 61, 57, 116, 129, 180, 182, 183, 186
    Cook calls on after Transit voyage: 92, 112, 118
Adventure: 116, *143*, 151, 173-4, 180, 183, 193
    leaves on expedition: 130
    death of men in New Zealand: 180-1, 183, 204-6
Aireyholme Farm: 32
Alaska: 215, 217, 235
Aleutian islands: 218, 219
Amherst, General Jeffrey: 42, 43, 44
Ancient Mariner: 14
Anderson, William (*Resolution*): 193, 218
Anson, Commodore George: 24, 33, 40
Antarctica: 132, 152
Antarctic Circle: 93, 136, 139, 141, 152, 153, 154, 186, 195
Antelope, HMS: 51
Arafura Sea: 22, 108
Arnold, John: 131, 141
Ascension Island: 26, 181
astrolabe: 125
astronauts: 18
Atlantic Ocean: 17, 21, 26, 52, 91, 178
Australia: 19, 22, 23, 25, 45, 91, 92, 93, 106, 108, 114, 173
Australia del Espíritu Santo: 22
    see also: Terra Australis

Bachstrom, Johann Friedrich: 37
Baltic Sea: 33
Banks, Sir Joseph: 61, 63, 64 *65*, 66, 68, 74, 76, 79, 80, 87, 89, 91, 92, 97, 112, 113, 117, 119, 126, 127, 128, 134, 182, 183, 186, *187*, 194, 231
    given place on *Endeavour*: 64
    background: 64, 117
    leaves *Resolution* for Iceland: 128
    relationship with Cook: 117, 128
Banks' Peninsula: 89
Batavia: 28, 76, 90, 109, 110, 111
Batts, Elizabeth *see* Cook, Elizabeth
Bay of Gaspé: 43
Bay of Islands: 87
Bayly, William: 131, 194, 201, 203, 216
Beaufort Sea (entrance to Northwest Passage): 219

Bering Sea: 218, 231
Bligh, William: 73, 193, 200, 202, 203, 214, 216, 218, 230
Board of Longitude: 73, 193, 200, 202, 203, 214, 216, 218, 230
Booby Island: 108, *109*
Bora Bora: 77, 164, 183, 211
Boscawen, Admiral: 41, 44, 45
Boswell, James: 183, 196
Botany Bay: 93, 96, *97*
Bougainville, Louis Antoine de: 25, 45, 75, 90, 129, 159, 169
Bouvet Island: 132
Bouvet de Lozier, Jean-Baptiste-Charles: 131, 132, 179
Buchan, Alexander: 61, 68, 69
Burney, James: 194, 198, 204-5, 216
Byron, Commodore John: 26-7, 62, 185, 186, 193

Canada: 34, 41, 43
cannibalism *see* Maori
Cape Circumcision: 131, 179
Cape Farewell: 91, 204
Cape of Good Hope: 23, 26, 27, 28, 91, 180, 194, 219
Cape Horn: 22, 23, 24, 27, 62, 67, 69, 91, 106, 140, 154, 175, 177
Cape Kidnappers: 84, 150
Cape Palliser: 141, 151
Cape Prince of Wales: 218
Cape Town: 22, 112, 128, 130, 132, 141, 178, 180, 181, 199
Cape Tribulation: 104
Cape Turnagain: 85
Cape Verde Islands: 200:
Carteret, Philip: 27-8, 185, 186
cartography: 51, 54, 88, 89 90, 97, 98, 116, 132, 139, 173, 202, 214, 216, 224
chronometer: 119, 121, 122, 125, 131, 141, 173, 200, 201
    Harrison's: 121-5, 131
Clerke, Charles: 189, 191, 192, 193, 194, 198-9, 210, 212, 216, 225, 228, 230, 231
    death of: 231
clothing: 188
Cocos Keeling: 13, *14*, 15
Colville, Lord: 49
compass, azimuth: 119
Cook, Elizabeth: 50, 51, 52, 182, 189, 198

after death of Cook: 231. 234
children: Elizabeth 118
George: 182
Hugh: 51, 198, 231, 234
James: 231, 234
Joseph: 118
Nathaniel: 52, 231
marriage to James: 50
with Cook on social occasions: 196
with Isaac Smith: 231
pension for, on death of Cook: 231
Cook, James
apprenticeship as merchant sailor: 32
birth of: 32
birthplace: 32
burial of: 230
career as sailor, chooses: 32
death of: 227, *229*
dinner party, accepts command third voyage: 192
education: 32
health of: 144, 155, 189, 196-8, 234
health of seamen, paper for Royal Society on: 188
ignored by Press after Transit voyage: 113
injuries to hand, Newfoundland: 52, 228
introduced to George III: 113
joins: *Endeavour*: 58
Freelove: 33
Grenville: 52
Pembroke: 41
Resolution: 116
Solebay: 41
journal on Great Southern Continent voyage; 186, 189
published: 190
Journal on Transit voyage: 113, 185
sees Hawksworth version of: 181, 185
to Bosun (promoted to): 36
to Commander: 112
to Lieutenant:58
Master : 41
Master's Mate: 35
to Post Captain: 186
portrait painted: 196
Royal Hospital for Seamen, Greenwich: 189, 191
Royal Society: admitted as Member of: 188
Copley Medal of: 188
views on Great Southern Continent: 153, 155, 175, 179

Cook, James (son of Captain): 51
Cook, James (father to Captain): 32
Cook Inlet (Gulf of Alaska): 218
Cook Islands: 147
Cook Strait: 153, 207
Cooktown: 12, 104
Coral Sea: 19, 22, 92
Crozet, Captain Julien Marie: 181

Dalrymple, Alexander: 28-29, 57, 58, 92, 113, 114, 119
refused command of Transit of Venus expedition: 113
Dampier, William: 25-6
Dance, Nathaniel: 127, 196
Deptford dockyards: 195, 196, 200, 224
Desolation Island: 175-6
Didycoy: 12, 15, 114, *114*, 120, 148
Diemen, Anthony van: 93
diet: 37, 64, 66,119, 138, 140, 142, 219
see also scurvy
Discovery , chosen for Northwest Passage voyage: 193
doldrums: 13, 14, *16*, 17, 130
Dolphin, HMS: 27, 28, 193
Drake, Sir Francis: 21-2, 23, 26
Drake's Passage: 21
Dusky Sound: 136, 139
Dutton: 181
dysentery: 110-1, 141

Eagle, HMS: 35, 36, 38, 39, 40, 41
Earl of Pembroke
see also *Endeavour*
Easter Island: 152, 155, 156, *160*
East Indies: 25, 27, 91, 92, 109, 118
Echo sounder: 15, 100
Elizabeth I: 20
Elliot, John (Resolution): 172
Ellis, William (*Discovery*): 194
Endeavour: 12. 18, 58, *59*, 60, 62, 63, 101, *101*, 186, 191
crew of: 61
damage to: 101
description of: 59, 60
repairs to (after striking reef): 97, *101*
strikes a reef: 101
Endeavour Reef: 104, 108
Endeavour River: 12, *97*, 104
English Channel: 63, 112
equator: 13, 19
Espíritu Santo: 172

Falkland Islands: 26, 27, 62, 116

'fearnought' clothing: 133, 212
Fijian Islands: 77, 208
fireships: 47, *47*, 48
Forster, George (*Resolution*): 129, 191
Forster, John Reinhold (*Resolution*): 128, 129, 131, 146, 149, 154, 155, 156, 163, 182, 189, 190, 194
France: 17, 34, 39, 40, 57, 195
Freelove: 33
Fresne, Marion du: 181, 202
Friendly Islands: 208
Furneaux, Tobias (*Adventure*): 139, 140, 141, 146, 148, 151, 152, 174, 180, 181, 183, 204, 206

George III, King: 29, 57, 97, 113, 125, 192, 231
Gibson, Samuel (*Resolution*): 117, 194
Gilbert, George: 116, 178, 210
Golden Hind: 21
Gonneville, Jean Binot de: 131
Gore, John (Discovery): 74, 193, 228
Graham, George: 122, 123
Great Ayton: 32, 186
Great Barrier Reef: 19, 22, 25, 92, 97, 100, 104, 107
Great Cyclades: 169
Great Southern Continent: 17, 18, 21, 22, 29, 57
    Admiralty instructions on Transit voyage to search for: 25, 62
    non-existence proved: 180, 186
Great Southern Continent expedition: 142
    Adventure crew attacked in New Zealand: 180-1, 183, 204-6
    Antarctic exploration in: 133, 134
    clothing of crew on: 133, 134
    health of crew (*Adventure*): 144
    message for Furneaux: 151-2, 173
Green, Charles: 58, 62, 68, 69, 74, 87, 108, 113
Grenville, HMS: 29, 54, 55, 56, 61
Gulf of Alaska (Cook Inlet): 218
Gulf of St Lawrence: 42

Halifax, Nova Scotia: 41, 42, 43, 45
Halley, Edmund: 29, 56, 121
Hamer, Captain: 36
Harrison, John: 121-5
Hartog, Dirck: 92
Hawaii: 196, 219, 220, 221, 235
Hawkesworth, Dr John: 185, 186, 189
health of the crew: 36, 38, 39, 64, 110-2, 119, 140, 141, 142, 144, 155, 156, 188
    see also diet; sickness; venereal disease:
Heights of Abraham: 48
Hodges, William (*Resolution*): 139, 140, 142, 162, 182
Holland, Lieutenant Samuel: 44
Huahine: 77, 146, 147, 162, 163, 211
Hudson Bay: 26

Ice into fresh water: 134
International Date Line: 12, 180
India: 34, 39
Indian Ocean: 15, 18, 23, 34, 91
Industrial Revolution:17
iron in trading, value of: 72-3, 145, 146, 212, 215

Jakarta *see* Batavia
Java: 23, 24, 109
Johnson, Dr: 183
Juan de Fuca Strait: 214-5

Kamchatka Peninsula: 231
Kauai: 212, 220, 221
Kealakekua Bay, Hawaii: 221, 222, 224, 226, 20, 235
    battle on beach: 226-7
    burial of Cook: 230
    death of Cook: 227, *229*
    reception at: 221-2, 224
Kendall, Larcum: 131
King, James: 193, 216, 222, 223, 225
King George's Land *see* Tahiti:
King George's Sound *see* Nootka Sound
King's Surveyor: 52
latitude, measuring: 71
    also see longitude
Lind, James: 37, 64
Linnaeus, Carl: 118, 131
Lizard Island: 107
longitude
    difficulties of measuring: 54, 69, 71, 119, 121
    lunar method: 120, 125
    see also chronometer
Lono: 222
Louisbourg: 42, 43, 44

Magellan, Ferdinand: 17-9
malaria: 11
Mangaia: 208
Maori: 79, 80, *81*, 84, 86, 88, 90, 149, 150, 174, 204
    cannibalism: 82, 181, 205

mapping *see* cartography
Marquesas Islands: 20, 62, 72, 77, 84, 156, 159
Maskelyne, Nevil: 125, 131
Matavai Bay: 72, 75, *143*, 144, 146, 160, 210
Maui Island: 220, 221
Mauritius: 131
Mendaña, Álvaro de: 20, 62, 156
Mexico, 19, 20
Mile End Road, Cook's home in: 51, 55, 118, 182, 198, 231
Miller, Frederick: 93
Monkhouse, William (*Endeavour*): 68, 110
Montagu, John *see* Sandwich, Earl of
Montcalm, General: 46, 48, 49
Moorea: 75, 210, 211, *211*
Murderers' Bay: 79

New Albion: 26, 195
New Caledonia: 173
Newfoundland: 25, 36, 37, 51, 56, 138
New Guinea: 19, 22, 26, 92, 107, 108, 115
New Hebrides: 22, 23, 25, 147, 169, 170, 172
New Holland: 23, 24
New Zealand: 62, 78, 84, 88, 90, 96, 118, 133, 135, 136, *137*, 140, 149, 194
    North Island: 79, 84, 85, 86, 141, 150, 152
    South Island: 89, 91, 136, 151
Nootka Sound: 215, 217
North America: 26, 34, 39, 40, 194, 195, 214
North Cape: 87
North Island *see* New Zealand
North Sea: 59, 116
Northumberland, HMS: 49, 50
Northwest Passage: 21, 26-7, 191, 195, 217, 218, 219, 230
    Prize for discovery of: 191
Northwest Passage expedition: 194, 195, 196, 230
    Cook's death on: 227, *229*
    Cook's erratic behaviour on: 200-1, 202-3, 204, 206, 207, 210, 226
    delayed schedule: 217
    retreat to Hawaii for winter: 219

Omai: 183, *184*, 189, 194, 203, 204, 211

Pacific Ocean: 13, 17, 18, 19, 21, 23, 26, 62, 92, 116
Palliser, Captain Hugh: 36, 39, 40, 41, 51, 52, 58, 191, 192, *197*, 198, 231
Parkinson, Sydney (*Endeavour*): 61, 80, 93, 98
Pembroke, HMS: 41, 43, 44, 45, 48
Peru: 20, 90
Philippines: 18, 19, 20, 113
Phillips, Molesworth (*Resolution*): 194, 222, 227
Pickersgill, Richard (*Endeavour*): 195 (*Resolution*): 136, 195
Pickersgill Harbour: 138
Pitt the Elder, William: 40, 45
Point Hicks: 93, 96
Portuguese seamen: 17, 19, 20
Poverty Bay: 79, 83, 84
Prince of Wales Island: 217
Princes Island: 111

quadrant: 74
Quebec: 40, 42, 45, 46, 50
    attack on: 47, 48
    fall of: 49
Queen Charlotte's Sound: 89, 93, 136, 139, 140, 150, 152, 173, 174, 183, 205
Queensland: 12, 19, 93, 96, 98, 105
Quirós, Pedro Fernández de: 20, 22, 164

Raiatea: 76, 77, 78, 162, 164, 183
Resolution on Great Southern Continent expedition: 116, 117, 132, *141*, *143*,
    alterations to: 118, 126, 127
    removal of alterations: 127
    Resolution on Northwest Passage expedition: 193, 200
    Cook's erratic behaviour: 200-1, 202-3, 204, 206, 207, 210, 226
    crew for: 193-4
    masts snap: 202
    relations with crew: 206-7
Rowe, murder of: 204, 206
Royal Society: 29, 54, 55, 56, 57, 58, 60, 61, 62, 77, 92

'Sailing Directions': 45, 49
St Helena: 181, 186
St Lawrence River: 41, 43, 45, 46, 49
St Pierre and Miquelon: 51
Sanderson, William: 32
Sandwich, Earl of: 52, 53, 118, 126, 147, 183, 190, 191, 192, 193, 198, 199, 218, 231
Sandwich Islands (Hawaii): 212
sauerkraut: 64, 119, 188

Savu: 108
Schouten, Willem: 23
scurvy: 24, 28, 36-8, 39, 42, 90, 119, 141, 142, 155, 156
    see also diet
seawater, freezing: 131
Seven Years' War: 39, 49
sextant: 125, *126*, 127
sexual relations with indigenous people: 73, 74, 145
Shadwell: 51
sickness at sea: 28
    see also diet, health, scurvy, venereal diseases
Siberia: 218
Silesia, conflict over control of: 39
Simcoe, Captain John: 41, 44, 49
Skottowe, John: 186
Skottowe, Thomas: 32, 186
Smith, Isaac: 231
Society Islands: 28, 77, 78
Solander, Dr Daniel Carl (Endeavour): 61, 63, 68, 69, 80, 87, 112
Solomon Islands: 20, 22, 23, 25
South America: 20, 21
South Georgia: 178
South Island *see* New Zealand
Spain: 20, 24, 39, 57, 66, 195
Spanish seamen: 17
Sparrman Anders (*Resolution*): 131, 144, 146
Spice Islands: 18, 21, 23
Spöring, Herman: 61
Staithes: 32
Staten Island: 67, 177, 178
stealing *see* thieving
Stephens, Phillip (Admiralty): 191, 231
Strait of Magellan: 17, 18, 20, 21, 23, 27, 28
Strait Le Maire: 67
Swallow: 27, 28
swimming ability: 214, 225
Sydney: 96

telescope, Gregorian: *56*
Tahiti: 25, 28, 29, 62, 70, 71, 72, 73, 75, 76, 77, 83, 90, 140, 142, 145, 152, 160, 161, *163*, 169, 189, 194, 207, 208
Tasman, Abel Janszoon: 79, 147, 164
Tasmania *see* Van Diemen's Land
Terra Australis Incognita: 17-8
    see also Great Southern Continent
thieving by indigenous people: 73, 74, 82, 146, 161, 163, 167, 171, 210, 212, 216, 225, 226
Tierra del Fuego: 17, 18, 21, 23, 67, 72
Timor: 24
Timor Sea: 22, 92
Tolaga Bay: 85
Tongan Group: 147, 148, *148*, 149, 164, 166, 208, 210
Torres, Luis de: 22, 92
Torres Strait: 22, 92, 108
Transit of Mercury: 87
Transit of Venus: 29, 56, 57, 74, 87
    location of observations for: 56, 57, 62
    setting up observatory for: 57, 70, 73, 75
    taking observations: 62, 113
Transit of Venus expedition: 57, 58, 92
    Cook's concern with diet and health: 111
    Cook selected to lead: 58
Triton: 39
Tuamotu Islands: 28, 92, 142, 159
tuberculosis: 199, 206, 218, 228
Tupia: 76, 77, 78, 80, 82, 83, 84, 110, 183

Unalaska Island: 218, 219, 220
Urdaneta, Andrés de: 19-20, 22

Vancouver, George (*Resolution*): 116
Vancouver Island: 214, 215
Van Diemen's Land: 93, 140, 202
Vanuatu: 22, 169
venereal disease: 25, 75, 213

Walker, brothers: 32
Walker, John: 32, 33, 189
Walker, Mrs (teacher): 32
Wallis, Commander Samuel: 27-8, 62, 72, 73, 74, 185, 186
watches: 124, 125, *125*
Webber, John: 194, 208, 221
Whitby: 32, 58, *59*
whitby colliers: 58, 116
Williamson, James (Resolution): 193, 212, 214, 226, 227
Woolfe, Brigadier: 42, 43, 44, 46, 47, 48, 49
Woolwich dock: 52, 55, 126

Yaws: 168

Picture Credits

A special word of thanks to the staff of The National Maritime Museum Picture Library in Greenwich, London, for their assistance with this book. (website: www.nmm.ac.uk). The author and Savitri Books would also like to thank David Jones for supplying the picture of the James Cook Memorial, in Anchorage, Alaska.

The publishers and the producers of this book would like to thank the following for providing photographs and for permission to reproduce copyright material. While every effort has been made to trace and acknowledge copyright holders, we would like to apologize, should there be errors or omissions.

Jacket front and back © The National Maritime Museum, London; endpapers, page 5 © Bill Finnis; 6 © The National Maritime Museum, London; 7, 10-11, 14, 16 © Bill Finnis; 18 © The British Library, London; 20. 30-1, 33 © Bill Finnis; 45, 47 © The National Maritime Museum, London; 49 © Bill Finnis; 53 © The National Maritime Museum, London; 55 © M Srivastava; 56 © The National Maritime Museum, London; 59 © National Library of Australia/Bridgeman Art Library, London; 65 © Agnew & Sons, London/Bridgeman Art Library, London; 70, 81 © The National Maritime Museum, London; 90 © Bill Finnis; 94-5 © The Natural History Museum, London; 96 © The National Maritime Museum, London; 99 © The Natural History Museum, London; 101 © The National Maritime Museum, London; 105 © M Srivastava; 109, 115 © Bill Finnis; 117, 123, 125 © The National Maritime Museum, London; 126 © Private Collection/ Bridgeman Art Library, London; 129 © Bill Finnis; 130, 133 © The National Maritime Museum, London; 138 © Bill Finnis; 141, 143 © The National Maritime Museum, London; 145 © Private Collection/The Stapleton Collection/Bridgeman Art Library, London; 148, 152, 157 © Bill Finnis; 158 © The National Maritime Museum, London; 160 © Bill Finnis; 163 © The National Maritime Museum, London; 165, 174, 182 © Bill Finnis; 184 © The National Maritime Museum, London; 187 © Lincolnshire County Council, Usher Gallery, Lincoln/ Bridgeman Art Library, London; 192 © M Srivastava; 197 © The National Maritime Museum, London; 199 © M Srivastava; 207, 209 © The National Maritime Museum, London; 211, 213 © Bill Finnis; 220, 223, 229 © The National Maritime Museum, London; 230 © Bill Finnis; 232-3 © The National Maritime Museum, London; 236 © David Jones; 243, 246.© Bill Finnis.